LEGO Technic Robotics

Mark Rollins

Apress·

LEGO Technic Robotics

ISBN 978-1-4302-4980-1

ISBN 978-1-4302-4981-8 (eBook)

President and Publisher: Paul Manning
Lead Editor: Michelle Lowman
Developmental Editor: James Markham
Technical Reviewer: Jurgen Krooshoop
Editorial Board: Steve Anglin, Mark Beckner, Ewan Buckingham, Gary Cornell, Louise Corrigan, Morgan Ertel, Jonathan Gennick, Jonathan Hassell, Robert Hutchinson, Michelle Lowman, James Markham, Matthew Moodie, Jeff Olson, Jeffrey Pepper, Douglas Pundick, Ben Renow-Clarke, Dominic Shakeshaft, Gwenan Spearing, Matt Wade, Tom Welsh
Coordinating Editor: Anamika Panchoo
Copy Editor: Mary Bearden
Compositor: SPi Global
Indexer: SPi Global
Artist: SPi Global
Cover Designer: Anna Ishchenko

Distributed to the book trade worldwide by Springer Science+Business Media New York, 233 Spring Street, 6th Floor, New York, NY 10013. Phone 1-800-SPRINGER, fax (201) 348-4505, e-mail orders-ny@springer-sbm.com, or visit www.springeronline.com. Apress Media, LLC is a California LLC and the sole member (owner) is Springer Science + Business Media Finance Inc (SSBM Finance Inc). SSBM Finance Inc is a Delaware corporation.

For information on translations, please e-mail rights@apress.com, or visit www.apress.com.

Apress and friends of ED books may be purchased in bulk for academic, corporate, or promotional use. eBook versions and licenses are also available for most titles. For more information, reference our Special Bulk Sales–eBook Licensing web page at www.apress.com/bulk-sales.

Any source code or other supplementary materials referenced by the author in this text is available to readers at www.apress.com. For detailed information about how to locate your book's source code, go to www.apress.com/source-code/.

This book is for any LEGO builder that dares to dream and then build what they can imagine. I would also like to dedicate this book to my wife, who showed me how to do a "photobox" that helped me to photograph my LEGO creations.

Contents at a Glance

Contents

About the Author

Mark Rollins was born in Seattle in 1971 and attended Washington State University in Pullman, Washington. He graduated in 1994 with a degree in English. After college, he began to write skits for college-age groups.

After four years working for Walmart and another five years working for Schweitzer Engineering Laboratories (SEL), Mark decided to pursue a full-time career in writing beginning in 2005.

Since then, he has written for many tech and gadget blogs, including screenhead.com, image-acquire.com, cybertheater.com, mobilewhack.com, carbuyersnotebook.com, gearlive.com, zmogo.com, gadgetell.com, gadgets-weblog.com, androidedge.com, and coolest-gadgets.com. He has also written for video game blogs such as gamertell.com and digitalbattle.com.

In 2009, Mark decided to create his own tech and gadget blog known as www.TheGeekChurch.com. The purpose of the blog was to report on the latest in technology, as well as inform the church-going crowd (who are often not very technically adept) of the benefits of using more technology in their ministry. Since 2012, Mark has devoted his time to this blog and considers it his ministry and mission.

Recently, Mark has become a Tech consultant, offering his years of experience in technology to consumer electronics companies.

Mark currently resides in Pullman, Washington, with his wife and three children.

About the Technical Reviewer

Jurgen Krooshoop is the founder of Jurgens Technic Corner (http://www.jurgenstechniccorner.com) and is the creator of many LEGO Technic MOCs (My Own Creation) and modifications for original sets. He's been creating these models for over 30 years. Often these models are motorized and remote controlled using Power Functions.

He has specialized in making complete step-by-step building instructions for Technic models. He created professional instructions for many of his models and models of other well-known builders.

He exposes his models at big LEGO events, such as LEGO-FanWelt in Cologne, and is an active member of LEGO forums such as Lowlug.nl and Eurobricks.com.

Acknowledgments

I would like to acknowledge Michelle Lowman, who approved of the book. I would also thank Mary Bearden, my copy editor, Jurgen Krooshoop, the techncial review editor, James Markham, who was very helpful in many areas, and finally, I would also like to thank Anamika Panchoo for all her hard work on this book.

Introduction

I am certain that everyone has a different image in mind when they hear the word "robot." One definition I found while doing a Google search stated that a robot was "a machine capable of carrying out a complex series of actions automatically." A good example from real life would be a mechanical arm on an assembly line designed to place specific parts on a product, but that would be all it would do. A second definition for robot is "a machine resembling a human being and able to replicate certain human movements and functions." These are the robots that we see in speculative fiction worlds that are mechanical creations made by a human to resemble a human or other organic creation.

The robots in this book are not the kind that have a computer "mind." Sadly, you will not be able to create a robot that can do your housework. These will be robotics that you can take control of with LEGO Power Functions and use for work or play, and you will find that they are capable of a lot of things. If you want programmable robots that will do things for you, I recommend *Mindstorms NXT*, which I mention later in this Introduction.

A Brief History of Robots

The idea of a robot stems from ancient civilization, and it is very interesting to see how our imagination has led to actual creations. Since their mental conception, robots have delighted us in fictional literature as well as in actual reality.

I'm not certain whether the goal of robotics is to create a mechanical creation indistinguishable from a human being, but I think it is interesting that we have never actually succeeded in creating one, even with our modern-day technology. The thing that I find most interesting is that societies that predate the first century A.D. tried to imagine a sentient automaton.

The idea of creating more than just an automated robot has thrived in literature, and the development of realistic robots and robots from fantasy went hand and hand. Some say that Mary Shelley's 1818 novel *Frankenstein* was essentially "the first robot story," even though it seems to imagine the monster (who is never referred to as "Frankenstein") as a flesh creation.

As literature inspired the creation of robots, robots began to inspire the idea of teaching a machine to essentially "think," which in turn led to the development of modern computing.

I am not certain what draws us to these machines. Perhaps it is that we humans enjoy creating, and want to take it to the next step and create something that actually has the ability to create.

Isaac Asimov coined the term "Frankenstein Complex" to describe the fear of mechanical men. I suppose that the takeover of a mechanical race is merely hypothetical with our present robotic technology. Fortunately, I can promise that none of the robots described in this book, which you can create, will ever try to kill you!

Why a LEGO Technic Book on Robots?

Looking at our history, both literal and literary, we, as humans, have been completely fascinated with robots. Even though we may not ever create a sentient one like Data from *Star Trek: The Next Generation*, we can't seem to shake the desire to do so. I would imagine that some of you are reading this book while iRobot's Roombas zip around your carpet and suck up excess debris. No doubt that you have other technological creations in your home that were originally constructed by robots working long hours on the assembly line.

LEGO Technic is most famous for its very realistic LEGO vehicles, but some of their sets focus on using Technic pieces to build some very interesting robotic creations. That changed in 2006, when LEGO introduced their LEGO Mindstorms NXT collection. I have tried out the Mindstorms NXT sets for myself and discovered that it is easier to make robots with pieces that seem specifically designed to be robot heads or arms and are completely programmable.

So I don't think the question is why write a book about how to create LEGO robots, but rather why write about how to create LEGO robots using nothing but LEGO Technic pieces and Power Functions. I don't feel any need to conceal that you could probably make robots easier and better with Mindstorms NXT kits, and Apress has several books on this subject. In the interests of full disclosure and displaying other good books from Apress, I recommend you go to www.apress.com and search for either LEGO or Mindstorms books.

If you want a Mindstorms book, you can find some there. If you want to, you could buy one of them, plus a Mindstorms kit, and start building. In this book I am focusing on LEGO Technic because I want you, the builder, to learn the basic building techniques so you can understand and perfect more advanced techniques. If you are reading this book, then you are a DIY person who doesn't mind doing some hands-on work. So yes, you could buy a toy robot that might do something better than the models I am describing here, but here you can actually learn how to make them. This is why I am going to spend a large part of Chapter 1 outlining how to use basic pieces from LEGO Technic. My goal as an author isn't that you just imitate the models in this book, but rather that you improve on them and create better robots.

LEGO Technic Robotics will teach you the basics, and understanding these basics of LEGO building will enable you to create better creations. You can then use the Mindstorms kits to simplify your work in the future.

How This Book Is Structured

I am not certain what level of LEGO builder you are. You may be just starting out and may never have put two LEGO pieces together. You might be someone who has been playing (or, if you prefer the adult term, building) with LEGO Technics for years. If you are the former, then this book is really going to help you out. If you are the latter, then you will enjoy creating the models in this book and discovering new techniques (or should I say Technics) for creating better creations.

- **Chapter 1: Where to Begin with Your LEGO Technic Robot Kit.** This chapter is as simple as it gets as I talk about basic Technic pieces and how they are used in creations. I also discuss how to purchase these pieces from popular sites like Brickfactory and LEGO's Pick a Brick, and I explain some techniques and tricks that are good to know when you work with them. I will also talk about software that will allow you to construct 3-D digital models of your LEGO Technic creations.

- **Chapter 2: Creating a Robot Body.** Using principles of geometry, this chapter talks about how to create a robot that can have squares, rectangles, and even triangles on its structure. This will be very helpful before creating robot arms and legs.

- **Chapter 3: Bring Your LEGO Technic Robots to Life with Power Functions.** I discuss the LEGO Power Functions and how they can bring LEGO Technics to life by allowing them to move. This chapter also explains how to create a wheeled base for a robot with four-wheeled steering, four-wheel drive, and suspension. Many of the concepts from my earlier LEGO Technic book are also reviewed here.

- **Chapter 4: Designing a Robot Arm.** Everyone wants a robot that can work its arms like a real person, and this chapter shows you how to create a robot limb that can bend at the shoulder, elbow, and wrist and even how to make an interesting artificial hand that can grip.

- **Chapter 5: Creating Robots with Extensions.** This chapter shows you how to make a robot with an extendable limb or waist. Here I detail two forms of extension with the rack and pinion and the scissorlift.

- **Chapter 6: The Robot Head.** This chapter discusses how to make a robot head and even how to give it expressions like eyes or a mouth. You may not be able to make it see, hear, and talk, but you can make it look like it does.

- **Chapter 7: Enabling a LEGO Technic Robot to Walk.** If you don't want your robot to roll around the floor on wheels, then you need to give it legs. This chapter shows you how to create a robot with two, four, or even more legs to make it walk.

- **Appendix A: Parts List.** This presents a complete list and description of all parts used in each project.

■ ■ ■

CHAPTER 1

Where to Begin with Your LEGO Technic Robot Kit

I'm sure you are anxious to get your hands on some LEGO Technic pieces and start building some robots. If you have worked with LEGO Technic before, you probably have a collection of all kinds of pieces from various sets accumulated throughout the years. I am going to assume that this is the case, but perhaps you are someone who wants to "start a LEGO Technic collection from scratch" or "from nothing". I'm trying to imply that the reader might be starting a LEGO Technic collection from nothing.

If you flipped through this book before you purchased it, then you will have seen a lot of instructions. If you want to try and build these LEGO creations yourself, you will obviously need to get specific LEGO Technic parts. There are several places online where you can do that, but it might be easier if you look at the Appendix in the back of this book. There you will find a list of all the parts you will need to build the individual models described in this book.

Before I start talking about where to purchase LEGO Technic pieces, let me say a few words about them. Traditional LEGO bricks are generally square and they come in brick form and flat bricks known as plates. They also have the studs to make them fit with the "female" side, but you will find that very few of the models in the book use studs at all. In their place are Beams, Levers, Axles, Connector Pegs, Bushes, Cross Blocks, Angle Elements, Gears, and various other Technic Miscellaneous Pieces.

LEGO Pick a Brick

The most obvious place to purchase LEGO Technic pieces is the LEGO web site itself. You can go to the official LEGO sitea t http://shop.LEGO.com. On the web site you will see under "Themes" the Pick a Brick section has a tab for "Categories," and here you can search for the LEGO Technic piece you are looking for under the Technic category. You can also do a search under Color Family. I know that some of you LEGO builders are sticklers for color and demand that your creations conform to a certain color scheme. You don't have that freedom if your LEGO collection is an amalgamation of many LEGO Technic sets over the years. In addition to searching by color, you can also search by its Brick Name, Element ID, or Design ID, which are specific designations that LEGO gives its parts.

If you are ever looking for a specific brick, you can do an Advanced Search on the left column using the Brick Name, which is the formal name for the brick. I will have to admit that will produce mixed results unless you know exactly what you are looking for. Later in this chapter, I will discuss some basic Technic LEGO pieces that you will need, and I will include the Element ID numbers.

When you are ready to purchase the part you are looking for, you can press the "Add To Bag" link and your individual pieces will appear in the "Brick Bag" column. When you have selected all the parts you need, just press the "Update Bag" and add the part to your personal Shopping Cart. Yes, you will need an account with LEGO to have your items sent to you, and you will be charged to that account upon checkout.

BrickLink

If you are looking for another place to find LEGO Technic pieces, I would suggest looking at BrickLink (http://www.bricklink.com), as shown in Figure 1-1. BrickLink is an unofficial LEGO marketplace, and it is often referred to as the "eBay of LEGO." If you want to buy or sell LEGO sets, new and used, this is the online place to shop.

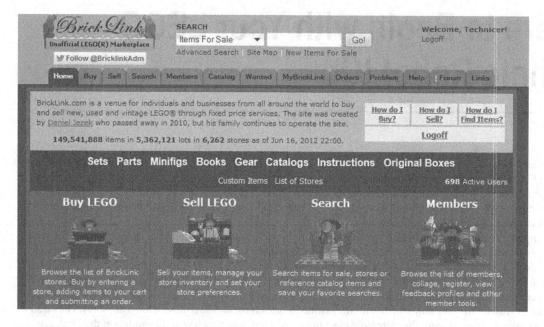

Figure 1-1. *The BrickLink site, a place to find LEGO bricks, both new and used*

If you click the Buy tab, you will have the option of purchasing several items, including sets, books, gear, catalogs, and parts. At the time of this writing, there are over 116 million parts available for purchase. Selecting the Parts tab will result in a category tree that branches out into several types of pieces, and there are 16 subcategories for Technic including:

- *Axle*: Anything that is an axle or has an axle attachment (see Figure 1-9).

- *Brick*: Any Technic shown in the Technic brick section in Figure 1-5, and some that I didn't show, is here.

- *Connector*: This is an umbrella term that refers to Angle elements and Cross Blocks (seeF igure1-22).

- *Disk*: These are disk-shaped pieces that I did not describe above. I don't really have them on any of the models in this book, and I don't really see them on more recent sets.

- *Figure Accessory*: At one point in time, Technic had figures that were to the scale of the Technic vehicles. They don't make them anymore, but here is a place where you can find the accessories like helmets and feet.

- *Flex Cable*: Some Technic sets have flexible cable that helps to create a more curvaceous shape. If that is what you are interested in, here is a place to find it.

- *Gear*: Parts like the one displayed in Figure 1-28.

- *Liftarm*: This refers to pieces like Beams and Levers, and all of their variations (seeF igures1-6 t o1- 8).

- *Liftarm Decorated*: This refers to pieces that have stickers or printed graphics on them.

- *Link*: A good example of a piece that will be used for steering on a LEGO Technic construction.

- *Panel*: These are very big pieces that take up a lot of space. I didn't really use any of them in any of my models in this book.

- *Panel, Decorated*. Like the other Panel pieces, but these often have stickers or some type of graphics on them.

- *Pin*: This is where you would find various types of Connector Pegs, which you can see in Figure1- 12.

- *Plate*: These are flat bricks with Technic holes in them. I didn't discuss them in this book at all and don't really have any models in this book that use them.

- *Shock Absorber*: These are the types of pieces that I discuss in Figure 1-37.

- *Steering*: These are various parts that would go well with steering.

Please note that the names and descriptions of parts that BrickLink might not be exactly the same as the names I have given to the parts in this book. I used the official names that LEGO designates their parts from a program they have known as LEGO Digital Designer (LDD), and these names can be different from what BrickLink calls them.

If you are looking to buy many LEGO Technic pieces, BrickLink is very similar to LEGO's Pick a Brick in that you can assemble your parts in a shopping cart and then checkout when you are ready. I found that their catalog is a little more extensive and easier to search through if you are looking for a specific piece, and you might be able to get a deal on pieces if you buy them in bulk. If you are looking to build one of the models in this book and want to purchase every piece for it, this is one place to go. It is even possible to purchase entire Technic LEGO sets that the company used to sell. In case you don't know, LEGO refurbishes their catalog every year, so their inventory is constantly changing.

Web Sites for LEGO Instructions

Some of you might want to build a LEGO Technic set that you remember making several years ago, but, as I stated before, LEGO changes their models every year. You might be able to find the actual set with instructions on BrickLink, and the more recent ones on LEGO.com, but if you have all the pieces, all you really need are the instructions. I highly recommend looking at the web sites listed below, just to generate ideas for LEGO Technic robots.

The LEGO Official Site

Oddly enough, every model that is available for purchase on the official LEGO Technic web site has a place where you can click and download instructions as a PDF file that can be saved on your computer. I have noticed that more recent LEGO instructions have a parts list in them, and this is a good information so you can order them on Pick a Brick or BrickLink. Keep in mind that LEGO likes to introduce new pieces as often as it produces new models. You may find that some of the newer pieces are harder to find, and you might have figured out how to build around it in some simple or complex workaround.

Peeron

If you are interested in building Technic LEGO sets over the years, then I highly recommend that you go to a web site that contains both LEGO catalogs and instructions. I found that Peeron (www.peeron.com) is especially helpful with its database of LEGO sets and catalogs (Figure 1-2).

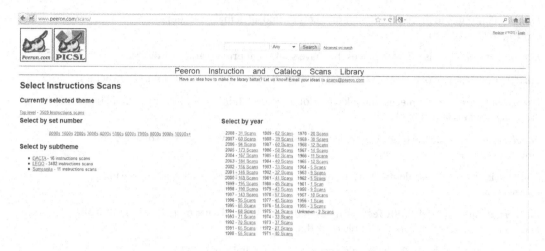

Figure 1-2. *A screenshot of the Peeron web site, a place to find instructions for LEGO sets*

I found that Peeron's inventory only goes up to the 2008 collections (as of this writing), and they often took a very long time to download. If you know the exact set you are looking for, then you should have no problem entering its name or ID number in the search engine and find its instructions and parts list. You might also be able to find a set if you know the date of its release. I will explain how to find Technic set numbers in the next section.

Brickfactory

Ia lsof oundB rickfactory(`www.brickfactory.info/`) to be a helpful resource, and it does have some of the more recent collections (Figure 1-3). Generally, LEGO Technic sets are given a number that is in the 8000 range or higher, with the exception of the 900 series when the Expert Sets first began in 1977. You will notice that several model series like Bionicle are filed under the same umbrella with advanced Technic sets.

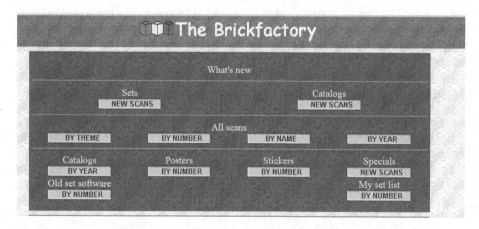

Figure 1-3. *The Brickfactory is a place to find all kinds of LEGO instructions*

You will discover that a lot of the models in Peeron and Brickfactory have most of their pieces available on Pick a Brick. Peeron is especially good at cataloging the individual pieces of a set, and you will find that a search for any set will reveal the individual pieces, including their individual Element ID. Unfortunately, you might discover that the Element ID on Peeron, BrickLink, or whatever site is a perfect match for the Element ID on Pick a Brick.

A Basic Introduction to LEGO Technic Pieces

There is no "ultimate set" of Technic pieces. The current selection in LEGO's catalog is for individual models, and some have many of one piece but not so many of another piece. If you are someone who isn't interested in spending too much of your budget on LEGO pieces, you learn to adapt your creations to the pieces you already have. In fact, you will discover that several of the creations in this book could probably be simplified with other LEGO Technic pieces, but I chose to stick to the basics for this.

A System of Storage for LEGO Technic Parts

As I said earlier, most LEGO enthusiasts simply build from whatever pieces they have based on the particular sets they have bought in the past. Before you do any building, I highly suggest you find a way to keep your LEGO Technic pieces organized. You will find you lose a lot of time rummaging through a pile looking for that one piece you need so you can move on to the next step.

As much as I like to hear the sound of LEGO bricks being scraped together, I like to keep my LEGO pieces organized in order to avoid long, drawn-out times of searching for "that right piece." I recommend buying some kind of tackle box, as the little drawers and storage containers on them are good for keeping pieces separate from one another (Figure 1-4). Another way is to purchase some kind of toolbox from a hardware or retail store that has drawers or other compartments for storing individual pieces. Of course you may not want this type of organization, and that is fine. The important thing is that you are have fun with this, and it does take quite a while to organize your LEGO pieces into sections like this.

Figure 1-4. *One way to organize LEGO Technic pieces, with tackle boxes*

Much of information in these next sections may seem elementary to most LEGO builders. I have already covered much of this in my first book, but I felt it necessary to explain the types of parts individually and their uses here. I will also show you how I organize my LEGO Technic pieces. You will note that in some illustrations, I don't have certain pieces, but I use an illustration from Peeron's LDraw just to show you what it looks like. See later in this chapter for more information about LDraw. Please keep in mind that the numbers on the illustrations are for the figure only, and I put the official LEGO ID number in parentheses so you can find it on Pick a Brick or BrickLink if the part is available for purchase.

This section discusses parts, but it is not meant to be an exhaustive list of what LEGO Technic parts are available to LEGO builders. As I mentioned earlier, LEGO creates new parts every year, and there are some parts that I deliberately didn't list here, as they were not used on any models in this book. I will also introduce other types of LEGO Technic parts in this book as I go, and will devote a lengthy section to it in the next chapter on Power Functions.

Occasionally, I will introduce a traditional LEGO brick into one of the instructions, but I didn't see a need to devote a section about what bricks and plates (flat LEGO pieces) to organize into drawers. If you want to, you could follow the pattern of organization in the illustrations below for your non-Technic bricks.

TechnicB ricks

When LEGO introduced their "Expert Sets" in the late 1970s, they introduced a new kind of brick. These bricks have holes on the sides for Connector Pegs and Axles (see individual descriptions below). Sometime around 2000, LEGO Technic began to emphasize Beams rather than traditional studded pieces. Most LEGO Technic sets shun these Technic brick pieces now and generally are completely studless. In fact, I can honestly say that I only used a handful of LEGO Technic Bricks in the projects in my previous book. However, if you are a LEGO builder of any type, you could devote an entire drawer in your tackle box or toolbox for these bricks, by length. As you can see, the amount of round holes in the side is usually equal to the amount of studs on top, minus one (with some exceptions). Figure 1-5 shows how I have this arranged in my tackle box, and you can see that I have a large space devoted to Technic Bricks of unusual shapes. You will note that I don't have them separated by color, and not every type of piece has a place by itself. This is because I have very few of some pieces and a lot of another, and it is just easier to group similar pieces together.

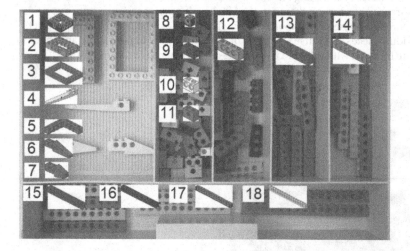

Figure 1-5. *A method of storing LEGO Technic Brick in a tackle box drawer*

■ **Note**   The numbers in Figures 1-5 through 1-9, 1-12, 1-22, 1-28, 1-37, and 1-38 correspond to the numbers in the parts list that immediately follows the respective figures. I put the official LEGO ID number in parentheses in the list so you can find it on Pick a Brick or BrickLink, if the part is available for purchase.

1. Brick,4 × 4 (32324). It is possible to make the same shape by assembling certain bricks together, but depending on your creation, it might be much easier to have this shape in one form. This is one of three fused quadrilateral shapes that LEGO has available.

2. Brick,4 × 6 (32531). Just like the 4 × 4 Brick, but slightly larger.

3. Brick,6 × 8 (32532). Also like the 4 × 4 Brick, and also larger.

4. Technic Fork (2823). This is essentially a 1 × 2 Technic Brick with an odd extension that takes it to a length of 6M. I think it is called a Fork because it used to be in LEGO sets with forklifts. It certainly resembles a basic prong of a standard factory forklift.

5. AngularB rick5 × 5 (32555). Like traditional LEGO Bricks, LEGO Technic Bricks also have corner pieces. Most of the traditional LEGO corner pieces are 2 × 2 bricks at a right angle, but this Technic model is 5 × 5 and will come in handy for all sorts of creations.

6. Wing Section, Rear (2744). A while back, LEGO Technic had sets with airplanes that used LEGO Technic Bricks. This was before Technic sets started using Beams (see next section), and this piece could create a Wing Section.

7. Wing Section, Front (2743). Like the Wing Section, Rear, this LEGO Technic Brick can form an excellent wing section in the form of a LEGO brick.

8. TechnicB rick,1 × 1 (6541). This is essentially a 1 × 1 LEGO Brick, but with a hole for an Axle or Connector Peg (see sections below). As stated before, the amount of round holes on the side of a LEGO Brick is generally equal to the number of studs, minus one. This is onee xception.

9. TechnicB rick,1 × 2 (3700). This is a 1 × 2 LEGO Brick, but with a single round hole in the middle.

10. TechnicB rick,1 × 2 with Two Holes (32000). This is like the 1 × 2 Technic Brick, but this one has two holes. It is one of two exceptions to the rule of studs and round holes, as explained earlier in this section.

11. TechnicB rick,1 × 2 with Cross Hole (32064). This is just like the 1 × 2 LEGO Brick, but instead of a hole for a Connector Peg or Axle, it has a Cross Hole made purposely to hold an axle in place (see part description below).

12. TechnicB rick,1 × 4 (3701). With one exception, LEGO Technic Bricks generally have an even number of studs, and this is one of the smallest at a length of 4M with three round holes on the side. The rest of the Technic Bricks increase by 2M or two studs, and the number of round holes on the side is equal to the number of studs, minus one.

13. TechnicB rick,1 ×6(3894).

14. TechnicB rick,1 ×8(3702).

15. TechnicB rick,1 ×10(2730).

16. TechnicB rick,1 ×12(3895).

17. TechnicB rick,1 ×14(32018).

18. TechnicB rick,1 ×16(3703).

Beams

As I mentioned earlier, the emphasis of recent LEGO Technic sets has been less about traditional LEGO pieces like Bricks and more about studless Beams. As you can see, organizing them is simple by length (Figure 1-6). Unlike Technic Bricks, the number of holes on the beam is equal to the length. Except for the 2M length, all beams tend to have an odd number of holes. Note that some of the Beams have a circled number beside them, and this is to indicate their length in terms of LEGO studs.

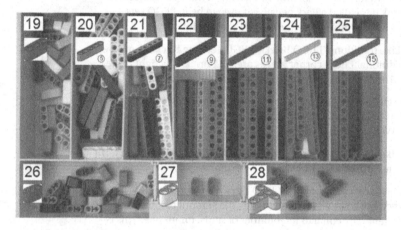

Figure 1-6. *A drawer in a tackle box filled with LEGO Beams*

19. Technic 3M Beam (32523). This is the beginning of the odd-numbered beams, which will increase by 2M in the next six drawers.

20. Technic5M Be am(32316).

21. Technic7M Be am(32524).

22. Technic9M Be am(40490).

23. Technic11M Be am(32525).

24. Technic13M Be am(41239).

25. Technic15M Be am(32278).

26. TechnicB eam1 × 2 Beam with Cross and Hole (60483). Also known as a Cross and Hole Beam, this has the unique ability to have a round hole for a Connector Peg or Axle and a cross-shaped hole made specifically for an axle. You will see this used in many creations in this book.

27. Technic 2M Beam (43857). As stated earlier, most Beams are odd numbered in their measurement. This 2M Beam is the only exception, save for the previously mentioned Cross and Hole Beam.

28. TechnicT -Beam3 × 3 (60484). This is essentially two 3M Beams fused together into a T Shape, and you will find that angled Beam pieces help on a lot of creations.

In addition to these straight beams, there are also beams at angles, and because of their number and size, I have devoted another drawer in the tackle box to them (Figure 1-7). In Chapter 2, I will discuss how to create triangular and trapezoidal creations with these Angular Beams.

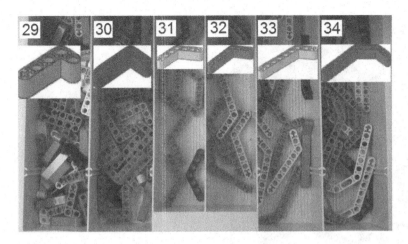

Figure 1-7. Another drawer for LEGO Beams

29. Technic Angular Beam 4 × 2 (32140). Like the T Beam, this is the merging of two Technic Beams to form a 4 × 2 Beam at a perfect right angle. This works well when creating square or rectangular creations. You will notice that one of the holes is cross-shaped, perfect for an axle.

30. Technic Angular Beam 5 × 3 (32526). Like the 4 × 2 Angular Beam, this is another 90-degree Angular Beam that is slightly larger at 5 × 3. Unlike the 4 × 2 Angular Beam, there is no cross-shaped hole on one end.

31. Technic Angular Beam 4 × 4 (32348). This Angular Beam is at an angle of 53.1 degrees. Note the cross holes at each end.

32. Technic Angular Beam 4 × 6 (6629). This Angular Beam is like the 4 × 4 as it is at the same angle at 53.1 degrees, but it measures at 4 ×6.

33. Technic Angular Beam 3 × 7 (32271). This Angular Beam has the same angle at 53.1 degrees, but measures at 3 ×7.

34. Technic Double Angular Beam 3 × 7 (32009). This particular Beam has two 45-degree angles, so it is essentially a 90-degree angle with a good curve to it.

Levers

Levers are essentially half the width of a Beam, and stacking two of them equals one Beam. Like the Beams, they often have Axle holes on the ends of them (Figure 1-8).

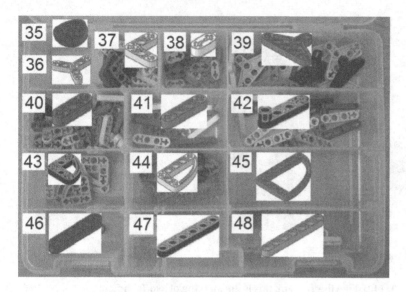

Figure 1-8. *Various LEGO Technic Levers*

35. Technic Comb Wheel (6575). This is not really a wheel in shape, but this lever is slightly round and it is good for securing things. I am guessing that it is called a Comb Wheel because it looks slightly like a comb.

36. TechnicL ever3 × 120 (44374). This propeller-shaped lever is called a 3 × 120 because there are three beams measuring 3M at 120-degree angles of one another.

37. TechnicL ever3 × 3M (32056). This is essentially two 3M Levers fused together at a 90-degree angle. Such a piece, with cross-shaped holes at its vertices, helps for all kinds of creations.

38. Technic 2M Lever (41677). This lever has two cross-shaped holes and is 2M in length, very helpful for securing Axles in place.

39. Technic Triangle (2905). This piece has an odd shape with five round holes and two cross-shapedh oles.

40. Technic 3M Lever (6632). This type of Lever has a cross-shaped hole on each end and a round hole in the middle.

41. Technic 4M Lever (32449). This type of lever is larger than the 3M, with two round holes in its middle and a cross-shaped hole on each end.

42. Technic 4M Lever with Notch (32006). The only difference between this piece and an ordinary 4M Lever is that the "notch" at the end is as thick as two stacked Levers (1M). You will find many places where this will work to your advantage.

43. Technic Half Beam Curve 3 × 3 (32449). This particular lever looks similar to the 3 × 3, but it has a circular curve joining two ends.

44. Technic Half Beam Curve 3 × 5 (32250). This is very similar to the Half Beam Curve 3 ×3, but it is larger at an elliptical curve at 3 ×5.

45. Technic Half Beam Curve 5 × 7 (32251). Like the Half Beam Curve 3 × 5, this also has an elliptical curve, but larger at 5 ×7.

46. Technic 5M Half Beam (32017). This Half Beam is aptly named as it looks like a 5M Beam split down the middle, with five round holes.

47. Technic 6M Half Beam (32063). This piece is just like the 5M Half Beam, but measures in at 6M. Since LEGO doesn't make any 6M Beams, you can also stack two of these atop each other and get the same effect.

48. Technic 7M Half Beam (32065). This is just like the 5M Half Beam, but measures in at 7M.

Axles

Like the Beams, the Axles can be easily organized by length (Figure 1-9). Axles are like bolts in the LEGO Technic world. You can use them to link just about anything, and their cross shape ensures that they fit on cross holes securely.

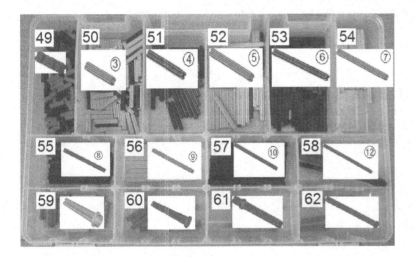

Figure 1-9. *A collection of LEGO Axles*

Like the Beams, the number in the circle is an indicator of the length.

49. Technic Axle 2M (32062). You can see in Figure 1-9 that 2M Axles comes in different colors, and I have noticed that newer ones come in a shade of red. Usually 2M Axles are notched, but I'm not certain why that is. Perhaps it is to make it easier for the LEGO builder to pry out.

50. Technic Axle 3M (4519). You can see in Figure 1-9 that some of the 3M Axles are black, but I have noticed newer sets using gray for the 3M. Generally, most Axles with an odd-numbered measurxement are gray in color.

51. Technic Axle 4M (3705). This is the second smallest of Axles with an even number in measurement. These even-numbered measured Axles are generally black in color.

52. Technic Axle 5M (32073). Like the 3M Axles, there are a few black 5M Axles in Figure 1-9, but these are usually gray.

53. TechnicAxle 6M (3706).

54. TechnicAxle 7M (44294).

55. TechnicAxle 8M (3707).

56. TechnicAxle 9M (60485).

57.   Technic Axle10M (3737).

58.   Technic Axle12M (3708).

59.   Technic Axle 3M with Knob (6587). Unlike ordinary 3M Axles, these have a bump on the end that can serve as a perfectly good LEGO stud.

60.   Technic Axle 4M with Stop (87083). Like the 3M Axle with Knob, the 4M with Stop has something that will make certain the Axle can only go so far. Although the stop is not a LEGO stud, I found these useful in a lot of projects, like when you need a 4M Axle to stayp ut.

61.   Technic Axle 5.5M with 1M Stop (59426). This piece is really 5.5M in length, and the stop is not placed at the end, but rather 1M from the end on one point. This is good for situations with wheel axles and other projects.

62.   Technic Axle 8M with Stop (55013). This is just like the 4M Axle with Stop, but twice aslon g.

Technic Axles in Action

In this book, I am going to show several LEGO models that you can build with the step-by-step instructions. I think it is worth my time to show you how some of these LEGO Technic parts work together. In Figure 1-10, you can see three 3M Axles with Studs on the left, followed by a 4M with Stop, a Technic Axle 5.5M with 1M Stop, and an 8M Axle with Stop.

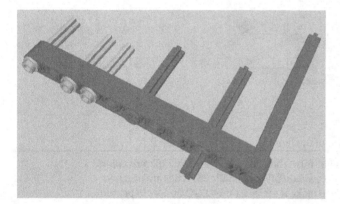

Figure 1-10. *A normal LEGO Technic 11M Beam with four types of Axles. Note how each type "caps" off a round hole in some way*

Each of the Axles in Figure 1-10 are stuck in such a way that they cannot be pulled out from their "uncapped" ends. You will find that in LEGO Technic creations, it is good to have Axles (not to mention other parts) secured on one end. Note the application of some other pieces in Figure 1-11.

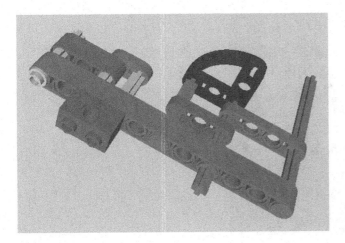

Figure 1-11. *This is the same structure as in Figure 1-10, but it shows how LEGO Technic Bricks, Beams, and Levers can fit on Axles*

You will note in Figure 1-11 that the two 3M Axles with Studs make it possible to stick on a 1 × 2 Technic LEGO Brick. It is possible to slide on the 3M Beam and 2M Lever. You will notice the varieties of Lever parts on the 4M Axle with Stud, 5.5M Axle with 1M Stop, and the 8M Axle with Stop.

In spite of these extra parts in Figure 1-11, none of these Axles are firmly held in place. For extra security in LEGO Technic, you can use both Connector Pegs and Bushes.

Bushes and Connector Pegs

If Axles are the nuts of the LEGO Technic world, then the Bushes are the bolts. The Bushes are a very common part in LEGO Technic builds, and they slide very easily on a Axle and are made to hold them in place. Like the other aforementioned parts, you will accrue a lot of these.

The Connector Pegs (sometimes called pins) are also very common parts, and they click into place in the round holes of the Beams or other LEGO Technic pieces. You will accrue a lot of Connector Pegs, and I consider them the "rivets" of the LEGO Technic world. There are other parts described in this section, and you can use them to link Beams and Levers together, and they come in many forms.

You will notice that Figure 1-12 is a sectioned-off container with these several types of parts, and you will see parts that look very similar but are different colors. The reason why I sectioned off the similar parts is because even though they look alike, they produce different effects. Some of the Connector Pegs have "friction," which means they do not spin as easily as the ones that do not have friction. You will discover that there are some times when you want a piece to spin freely and easily, so you will then want the pieces that do not have friction. Then there are times where you want a construction to lock securely in place, so using a piece with friction is your best course of action. I will discuss more about specifics of construction later in this chapter and other chapters, but here I only wanted to call it to your attention.

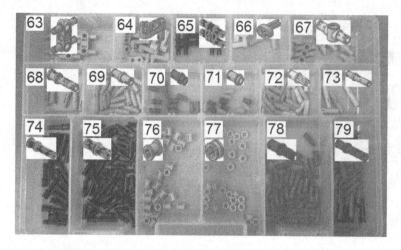

Figure 1-12. *A collection of Technic Connector Pegs and Bushes*

63. Technic Beam 3M with Four Snaps (48989). I'm not certain just how to describe this, but it has four connector pegs, with round holes on another and a round hole in between the connector pegs. It is very useful for securely locking something with a round hole together.

64. Technic Steering Knuckle Arm 2 × 1 × 3 (33299). This is like a Lever with half a Connector Peg on one side and a cross-hole notch on the other. I believe it is called a Steering Knuckle Arm because it comes in handy for constructions that require steering.

65. Technic Module Bush (32138). This is similar to the 3M Beam with Four Snaps. It has two Connector Pegs on each side, but no 1M space in the middle. You can insert an Axle in the middle via the cross-shaped hole.

66. Technic Plastic Motor Crank/Cross (2853). If you insert a 2M Axle into one side of a 2M Lever, you will essentially have this part. It is often used for motors, but they have other uses as well.

67. Double Bush 3M (87082). This is essentially a Connector Peg 3M in length with a round hole in the middle.

68. Technic Friction Snap with Cross Hole (32054). As I mentioned earlier, some pieces have friction and some do not. I thought that the Friction Snaps had looser pieces, but I don't think they do any more. Perhaps older sets have this, as I seem to have a handful in my own LEGO Technic collection that spin quite freely. Normally, the part that has friction has a different part number, but I could not find any alternative part number for this one.

69. Technic Connector Peg (3673). This piece is designed to snap into a round hole and the other will snap into another hole.

70. Technic 1 1/2 Connecting Bush (32002). This piece is three-fourths the size of a Connector Peg, with a bump half the size on one side. The half-sized bump is good for securing a round hole of a Lever.

71. Technic Connector Peg with Knob (4274). Similar to the 1 1/2 Connecting Bush, it has one side that is a Connector Peg and the other side a knob that is a hollow LEGO stud.

72. Technic Connector Peg with Cross Axle (6562). One side is a Connector Peg, the other side a 1M Axle. This piece can link up a part with a round hole and a part with a cross hole very securely, and still allow for some spinning.

73. Technic Connector Peg 3M (32556). This Connector Peg can hold two 1M pieces on one side, so it can join three Beams together so they can freely spin.

74. Technic Friction Snap with Cross Hole (32054). The Connector Peg that you see here has some friction on it so you can lock two things together securely, and they will not spin toof reely.

75. Technic Connector Peg with Friction (2780). Like the other type of Connector Peg, this has some friction going on. I would have to say that this is the part, along with the Bushes, are the parts that I use the most with LEGO Technic creations, as it is the easiest way to link two parts with round holes together.

76. Technic Bush (6590). This piece is about 1M in length, and fits snug on an Axle. It is made to hold an Axle in place and has many other uses.

77. Technic Half Bush (32123). The Half Bush is only 1/2 M in length and has the same function as the Technic Bush.

78. Technic Connector Peg with Friction and Cross Axle (43093). This is the Connector Peg/Cross Axle with Friction.

79. Technic Connector Peg 3M with Friction (6558). This Connector Peg can hold two 1M pieces on one side, so it can join three Beams together so they cannot freely spin.

Let's look at how these parts work with other LEGO Technic parts. I will begin with the Connector Pegs, those with friction and those without.

Technic Connector Pegs in Action

Figure 1-13 shows a 15 Beam with some Connector Pegs and other beams attached. On the left is a Connector Peg with Friction, and one peg is enough to allow the 5M Beam to turn 360 degrees. Since it has some friction, it won't spin like a well-oiled machine, and it has a limited degree of holding its angle. It is good for when you want a part of your project to turn, but not too freely.

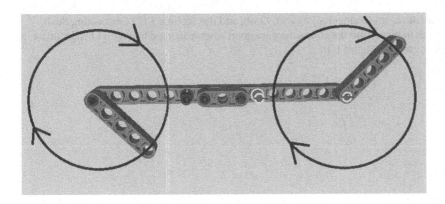

Figure 1-13. *A 15M Beam connected to a 3M Beam and two 5M Beams with two types of Connector Pegs*

In the center of Figure 1-13, you will see both types of Connector Pegs on the left and right sides of a 3M Beam. You will note that the 3M Beam is held securely by two Connector Pegs. You can see that two Connector Pegs are all that is required to securely hold two Beams together.

On the right of Figure 1-13, a 5M Beam is held in one place by a regular Connector Peg. This one will spin like a well-oiled machine and is good when you want a part of your project to spin freely, like a wheel.

Connector Pegs also come in 3M forms, which come in handy for linking three parts together. You can see in Figure 1-14 how they work.

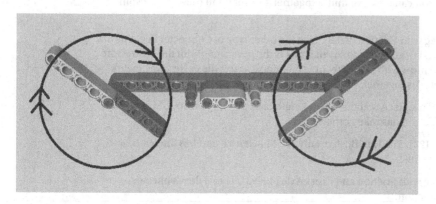

Figure 1-14. *Similar to Figure 1-13, this is a 15M Beam linked to 3M and 5M Beams with two types of 3M Connector Pegs*

This time, I will start with the center of Figure 1-14. You will note that there are two of each type of 3M Connector Pegs in the center of the 15M Beam. You will notice that there is a visible ring on one end, with enough space to push the peg in 2M on one side and 1M on the other side. You can see that in the exact center it is quite easy to stack three Beams together.

You will note that on the left of Figure 1-14 there are two 5M Beams held in place by a 3M Connector Peg with Friction. Both Beams can spin freely and have some degree of holding in place.

On the right of Figure 1-14, these two Beams will spin quite freely.

Technic Bushes in Action

Now I will focus on the Bush, the Half Bush, the Connector Peg with Knob, and the Technic 1 1/2 Connecting Bush. The bushes are good for securing axles in place, and the connectors are good when you need to have a LEGO stud as part of your LEGO Technic creation, as seen in Figure 1-15.

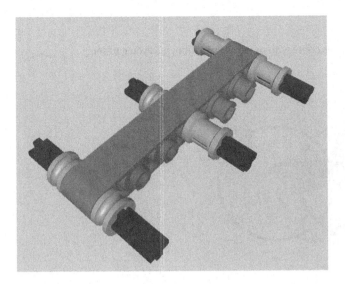

Figure 1-15. *A normal 7M Beam with three Axles, secured with Bushes and Half Bushes. Note the placement of the Connector Peg with Knobs and 1 1/2 Connecting Bushes*

You will notice that all the Axles in Figure 1-15 are secured on the Beam by Bushes, Half Bushes, or a mixture of both. This will hold the Axles in place securely, but they will still spin in the round hole. As for the Connector Pegs with Knobs and the Technic 1 1/2 Connecting Bush, see Figure 1-16 for what you can do with them.

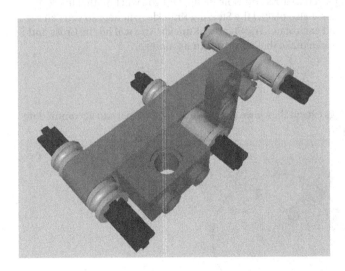

Figure 1-16. *This is the same construction as Figure 1-15, with a Brick placed on the Connector Pegs with Knobs. Note how a Lever with a round hole can be attached on the 1 1/2 Connecting Bush*

Note in Figure 1-16 that the 1 × 2 Technic Brick is secured in place on the two Connector Pegs with Knobs that are very visible in Figure 1-15. You can see that a 3M Lever is placed on the Technic 1 1/2 Connecting Bush.

Connector Peg/Cross Axles in Action

Like the Connector Pegs, the Connector Peg/Cross Axles come with or without friction, and you can see a demonstration of them in Figure 1-17.

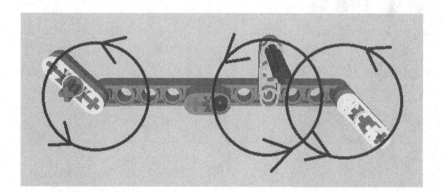

Figure 1-17. *A normal 11M Beam with two types of Connector Peg/Cross Axles, as well as a Steering Knuckle Arm, Connector Peg, 4M Axle, Cross and Hole Beam, and four 3M Levers*

You will notice in Figure 1-17 that I stacked two 3M Levers on each side of the 11M Beam and secured them with two types of Connector Peg/Cross Axles. As I mentioned earlier, two stacked Levers equal the width of one Beam. You can even see that I have placed a Connector Peg/Cross Axle in the center of the 3M Levers.

In the center of Figure 1-17, you will notice that I placed a Cross and Hole Beam, and you will find that this 2M piece comes in handy in a lot of the creations in this book. I also placed the Steering Knuckle Arm here so you can see how it resembles the sides with the 3M Levers and how it can rotate. The Steering Knuckle Arm will hit the Cross and Hole Beam if spun at certain angles and sometimes you can make that work to your advantage.

Technic Friction Snaps in Action

The Friction Snaps look a lot like the 3M Connector Pegs except they have a cross-hole section that can accommodate 1M of an Axle. You can see them at work in Figure 1-18.

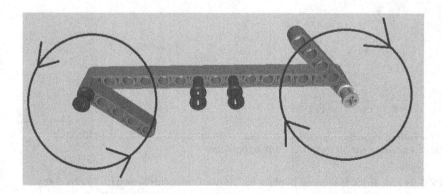

Figure 1-18. *This is a 15M Beam with a 5M Beam on each side, secured by two types of Friction Snap*

You can see in Figure 1-18 that each type of Friction Snap allows for spinning movement of the 5M Beams. As mentioned earlier, there are no Friction Snaps without friction, at least not in recent Technic sets.

In Figure 1-19, I show several things you can do with Friction Snaps and other parts.

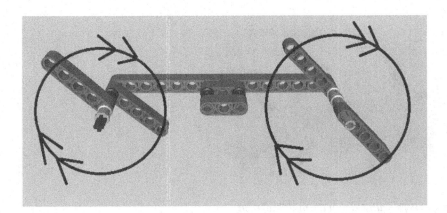

Figure 1-19. *This is essentially the same construction as in Figure 1-18, but I have added more pieces to it*

On the left side of Figure 1-19, there is a 4M Axle stuck into the Friction Snap, and then a 5M Beam with a Bush is put on.

In the center of Figure 1-19, two Connector Peg/Cross Axles are placed on the Friction Snaps, and then a 3M Beam is snapped on. There is also another 3M Beam put on below the cross-hole ends of the Friction Snaps.

On the right of Figure 1-19, a 2M Axle has been placed on the Friction Snap, and another Friction Snap has been put on top of that. A 5M Beam is then slid on this second Friction Snap.

Other Types of Technic Connectors in Action

This section will focus on three more parts: the Double Bush 3M, the Module Bush, and the Technic Beam 3M with Four Snaps. You can see them in order from left to right in Figure 1-20 (with two variations of the Double Bush 3M).

Figure 1-20. *A normal 15M Beam with three types of Connector Pegs*

You will notice that in Figure 1-21 I have placed another 15M Beam to create an interesting "sandwich" with the Connector Pegs from Figure 1-20 inside. You will also notice that I have put Axles through the round holes and secured them with Bush and Half Bushes. The Axle in the Module Bush did not need any Bushes to secure it as it has ac ross-shapedh ole.

Figure 1-21. *Three types of Connector Pegs, with Axles in the middle of them*

Cross Blocks and Angle Elements

You are going to find it necessary to link beams together in a perpendicular manner. In other words, you are going to have a part with holes facing one direction, but you often want to make it so the holes face 90 degrees in the other direction. You can shift angles in many creative ways using the nine types of Cross Blocks that you see in Figure 1-22. As for the Angle Elements, you can see six examples of them in Figure 1-22, and these pieces are designed to link Axles together so they fit at certain angles.

Figure 1-22. *A collection of Cross Blocks and Angle Elements*

80. Technic Cross Block 90 Degrees (6536). This is a part 2M in length with a round hole in one direction and a cross hole facing 90 degrees the other way.

81. Technic Cross Block/Fork 2 × 2 (41678). This part is two round holes and then a cross hole on each side, at 90 degrees below it.

82. Technic Cross Block 2 × 3 (32557). This piece is interesting with two round holes in one direction and two round holes below it at 90 degrees and centered.

83. Technic Double Cross Block (32184). I found this piece to be very useful to join together some pieces. It has two cross holes on two sides, with a round hole on two other sides.

84. Technic Cross Block 3M (42003). This is a 3M Beam with two round holes and a cross hole on 90 degrees on the other side.

85. Technic Cross Block Form 2 × 2 × 2 (92907). This piece as a round hole, and then four cross holes (two on each side) 1M above it, faced 90 degrees away.

86. Technic Cross Block 3 × 2 (63869). This is like having a 3M Beam, and then gluing a Bush on the top and center, and turning it 90 degrees.

87. Technic Steering Gear (32068). This Cross Block is, as its name implies, used for steering, but it also comes in handy for other uses as well.

88. Technic Cross Block 1 × 2 (32291). This is a lot like the Cross Block 3 × 2, but it is a Bush mounted 90 degrees on a 2M Beam instead of a 3M Beam.

89. Technic Zero Degree Angle Element #1 (32013). Used for capping off an Axle with a round hole at 90 degrees.

90. Technic 180 Degree Angle Element #2 (32034). Used for joining two Axles together in a straight line, with a round hole in the middle turned at 90 degrees that can also be used for an Axle.

91. Technic 157.5 Degree Angle Element #3 (32016). This Angle Element and the three others in this section are made to join two Axles together, with the round hole in the middle turned at ninety degrees for more uses. The name of the Angle Element is equivalent to the anglei ti s.

92. Technic 135 Degree Angle Element #4 (32192).

93. Technic 112.5 Degree Angle Element #5 (32015).

94. Technic90D egreeA ngleE lement# 6(32013).

Technic Cross Blocks in Action

There are several ways to use Cross Block and Angle Elements, and I want to highlight some of the more common ways, starting with the Cross Blocks. I will start with the Double Cross Block, the Cross Block 3M, and the 90 Degree Cross Block, which you can see in Figure 1-23.

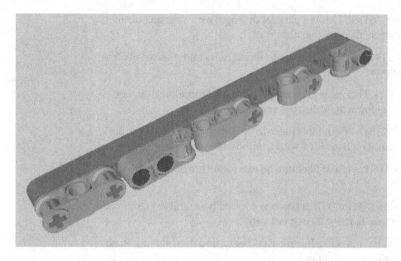

Figure 1-23. *A normal 15M Beam with a Double Cross Block, two 3M Cross Blocks, and two 90 Degree Cross Blocks. Note the ways they are attached to the Beam, by Connector Pegs and Connector Peg/Cross Axles*

You will notice that the 15M Beam in Figure 1-23 has a Double Cross Block held in place by two Connector Peg/Cross Axles. On the right of that is a Cross Block 3M held in place with two Connector Pegs. These Cross Blocks are securely held in.

The other three Cross Blocks will spin freely (until their rotation is blocked by other LEGO parts), but if Connector Pegs and a Beam are placed on them, as in Figure 1-24, they will be quite secure.

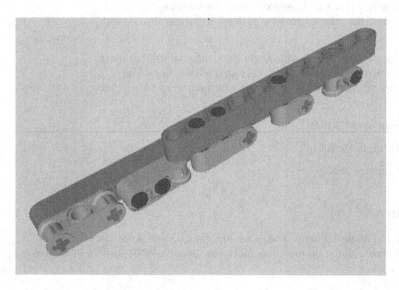

Figure 1-24. *Once the Cross Blocks are in place, a Beam or other LEGO Technic pieces can be secured at 90 degrees*

Not only does the addition of the 11M Beam make certain that the three Cross Blocks stay aligned, but it also allows the builder access to round holes on top of the structure, rather than just the round holes on the sides. You will discover that in building you will constantly have to change sides to add more and more parts, and Cross Blocks are the best way to do this.

Three other types of Cross Blocks are the Cross Block 3 × 2, Cross Block 2 × 1, and the Cross Block 2 × 3, all of which are shown from left to right in Figure 1-25.

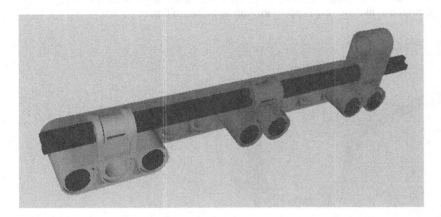

Figure 1-25. *Three more types of Cross Blocks attached to an 11M Beam. Note how the 12M Axle is able to link them together securely*

The one thing that you will notice is that the Cross Block 3 × 2 has a cross hole that extends in the center, so it lines up with the round hole on its base. The 2 × 1 and 2 × 3 Cross Blocks have sections that are centered in the middle of two round holes. This means that you are going to have round holes in the center of two round holes if you choose to use this method.

Another type of Cross Blocks has a "fork" formation. This includes the Cross Block Form 2 × 2 × 2 and the Cross Block Fork 2 × 2. You can see them from left to right in Figure 1-26.

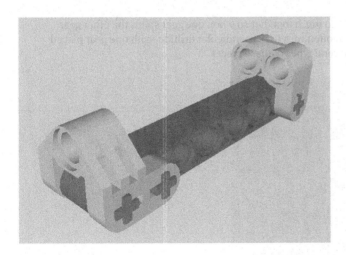

Figure 1-26. *Two types of Cross Blocks on a 7M Beam, attached via 2M Axles*

You will note that these are held in place with 2M Axles, that the Cross Block Fork 2 × 2 has two holes, and they are not aligned with the holes on the 7M Beam below it.

Technic Angle Elements in Action

The Angle Elements are used to link up Axles at varying angles. By using many together, like Angle Elements 1 through 6, you can make strange formations like the "Big Dipper" formation shown in Figure 1-27.

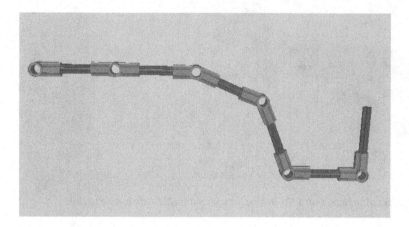

Figure 1-27. *You can see the six Angle Elements used together here. If you were to use these in real life, you will see different numbers on the parts that mean different angles*

In Chapter 6, I will demonstrate how many of one type of Angle Element can be combined to create almost circulars hapes.

Gears

Gears are designed to spin, and the teeth are designed to mesh together so when one gear spins, the other gear spins in turn. They come in various sizes, and they can often turn in perpendicular fashion, with one gear placed at a 90-degree angle to the other. Figure 1-28 shows several types of these pieces.

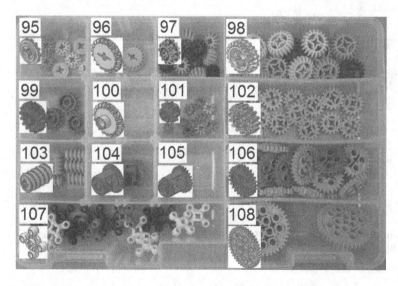

Figure 1-28. *A way of organizing gears for building your Technic projects*

95. Technic Conical Wheel Z12 (6589). This gear is flat with 12 teeth, and it can spin in a perpendicular fashion with another wheel.

96. Technic Bevel Gear Z20 (32198). Similar to the Conical Wheel Z12, this gear is slightly larger at 20 teeth and can also spin in a perpendicular fashion.

97. Technic Double Conical Wheel 12 Teeth (32270). Like the Conical Wheels, these can spin in a perpendicular fashion, but they are also thicker.

98. Technic Double Conical Wheel 20 Teeth (32269). This is like the previously mentioned Double Conical Wheel, but slightly bigger.

99. Technic Gear Wheel (16 Teeth) Special (6542). Instead of having a cross hole in the center, this has a round hole. It is helpful for situations where you need another gear, such as shifting gears on a LEGO Technic vehicle.

100. Technic Cone Wheel (87407). I think this gear gets its name from being slightly conical in shape as the area about its round hole (not a cross hole) extends a bit.

101. Technic Gear Wheel 8 Teeth (3647). This is one of the smallest gear pieces, and it has a cross hole in the middle.

102. Technic Gear Wheel 16 Teeth (4019). This gear is slightly bigger than the 8 Teeth, with a single cross hole in the middle.

103. Technic Worm Gear (4716). This gear is designed to mesh with another round gear above it. By turning this Worm Gear, the gear connected to it will turn.

104. Technic Differential Gear Casing (62821). This piece allows wheels to spin better on a properly made axle, and this particular one requires three Z12 Conical Wheels to work properly.

105. Technic Differential 3M (6573). This is similar to the other Differential Gear Casing, with an extra gear on it.

106. Technic Gear Wheel 24 Teeth (3648). This is similar to the other Gear Wheels, but it has three cross holes and four round holes.

107. Technic Angular Wheel (32072). I found that these Angular Wheels mesh together and spin well in a perpendicular fashion.

108. Technic Gear Wheel 40 Teeth (3649). This is quite a huge LEGO piece, and it has 12 round holesa ndf ivec rossh oles.

LEGO Technic Gears in Action

Working with Gears is essential for making certain Technic creations work, and it is important that you understand how to properly orient them. Here are five rules about gears that you need to know:

1. *The gears must mesh together, perfectly.* Gears that are not secure tend to shake loose and often jam. If you secure them with the Axle holding the Gear going through at least 2M of Beam or other Technic part, it usually works well.

2. *If one gear turns, the other gear will turn in the opposite direction.* The rules of mechanical engineering apply to LEGO, and if one spins clockwise, then the other will spin counter-clockwise. If you intend to link many gears together, you might want to make certain that they spin in the direction you want them to.

3. *Do not use too many Gears in one construction.* I have discovered that using many Gears in parallel or perpendicular fashion is not wise and at times can cause jamming, even with proper securing. I don't have an exact number, but you might want to consider alternatives to many gears in a formation.

4. *Some gears will not mesh together.* I explained how certain Gears mesh together, but not all of them will do so perfectly and some not at all. In order to make certain your construction works well, use Gears designed to mesh well together.

5. *Gears will always mesh at different angles.* You cannot mesh two gears together without one of them being turned at a slight angle so the teeth can intersect. For example, the Angular Wheels have to be at 45 degrees with each other in parallel or perpendicular construction.

Let's start with the 8-, 24-, and 40-tooth gear. Each of these can be put in a series, as seen in Figure 1-29.

Figure 1-29. *I have lined up three types of Gears (the 8-tooth, 24-tooth, and the 40-tooth Gears) so you can see the space in between*

In case you don't see a pattern in Figure 1-29, count the round holes in between the gears.

- 0Mb etweent wo8 -toothG ears

- 1M between an 8-tooth and a 24-tooth Gear

- 2M between two 24-tooth Gears

- 3M between a 24-tooth Gear and 40-tooth Gear

- 4Mb etweent wo4 0-toothG ears

Another type of gear has 16 teeth, and they can be spaced apart with one hole in the middle, as seen in Figure 1-30.

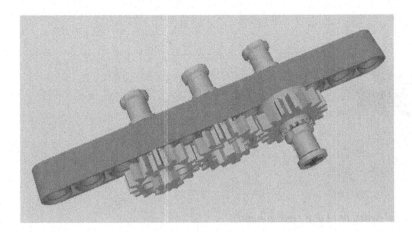

Figure 1-30. How to orient 16-tooth Gears to they mesh perfectly on a Beam

You will notice that the Gear all the way to the right in Figure 1-30 is a special Z16, with a round hole in the middle instead of a cross hole. This is why it is secured in place with an additional Bush.

The Double Bevel or Conical Gears, such as the Z12 and Z20, do not mesh well when placed together individually with each other. But when the Z12 and Z20 are placed together with once space apart, they mesh perfectly, as seen inF igure1-31.

Figure 1-31. How to mesh Conical Gears together—alternate the Z12 with Z20

While I was working with the Z12 Gears, I noticed that they would mesh diagonally, as seen in Figure 1-32. It is an unofficial gear combination and could have too much friction.

Figure 1-32. *A way to mesh two Z12 Conical Gears, but I don't recommend it*

I found that other LEGO builders did not recommend this type of construction, so I don't really use it unless there is no other way around it. Interestingly enough, the Double Bevel Gear is able to mesh together in a perpendicular fashion, a ss eeni nF igure1- 33.

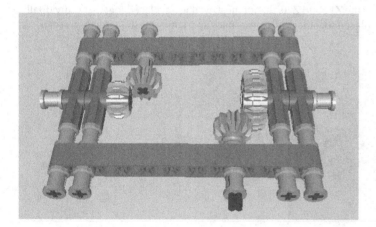

Figure 1-33. *In this formation, you can spin a Conical Gear, and the other meshed at 90 degrees from it will spin as well*

You will notice that the Z12 and Z20 Conical Gears in Figure 1-33 mesh together, and the two Z12 Conical Gears also mesh together. Notice how long the Axles are on each. If this were a real project, I would recommend two layers on each Axle so it and the Gear are secured.

The Single Bevel Gears can also function in a perpendicular capacity. In fact, if they are placed in a Differential, they make a perfect axle for a vehicle (see Figure 1-34), and this is a popular design for many LEGO vehicles.

Figure 1-34. A LEGO Differential, with three Single Bevel Z12 Gears. If the larger Gear on the Differential is meshed with another gear, you will have a good axle for a wheeled vehicle

Speaking of perpendicular ways of Gear construction, I highly recommend the Angular Wheel. This X-shaped oddity only requires a space apart of 1M, and it can be turned so it works in a perpendicular fashion. If you look closely at Figure 1-35, you can see how they mesh together.

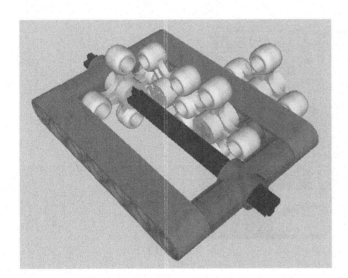

Figure 1-35. If any one of the Angular Wheels turns in this construction, the one on the end will also, thanks to both parallel and perpendicular building. Note how each "plus" form meshes with an "X" form

Another interesting gear formation is the Worm Gear, and it is essentially a way of turning a threaded tube (the Worm Gear) so the circular gear above it can rotate. This is a very effective use of gears, and the best part about it is the circular gear is locked in place when the Worm Gear stops turning. You will see the Worm Gear used many times in this book. You can see an example of Worm Gear constructions in Figure 1-36.

Figure 1-36. *These are two examples of constructions that use Worm Gears. If the Z16 Conical Gear is spun, then the circular gears will spin and so will the Angular Wheel. Note the one on the left is a 24-tooth Gear and the one on the right is an 8-tooth Gear. Notice the difference in individual construction*

Racks and Shock Absorbers

The Rack pieces are designed to work with gears, and they can be used with a gear so they can shift in ways that will create some interesting creations. You will also discover that the Shock Absorber pieces work well when you want to create a wheeled vehicle that can handle tougher terrains, as they give the frame some bounce to it.

I realize that I showed examples of how to use LEGO Technic pieces in previous sections, but I did not in this case as these types of constructions are quite complicated. I do use these pieces later in this book and will discuss them when the time comes. They are displayed in Figure 1-37.

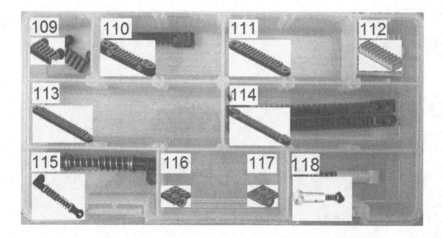

Figure 1-37. *A selection of Racks and Shock Absorbers*

109. Technic Rack with Ball (6574). This is a small rack with two balls on it that helps in steering functions.

110. Technic Rack 7M (87761). This is a rack with a round hole on each end, with a 5M section of "teeth" in the middle. It also has two cross holes on the side.

111. Technic Toothed Bar 8M (6630). This particular rack doesn't have a round hole to allow for Connector Pegs, but it does allow for studs.

112. Technic Toothed Bar 1 × 4 (3743). This is designed to connect to a 1 × 4 row of studs and good for rack-and-pinion steering.

113.   Technic Toothed Bar 10M (6592). This is like the 8M Toothed Bar, but longer.

114.   Technic Rack 13M (64781). Like the 7M Rack, it has the round holes for Connector Pegs and the Axle holes on the side.

115.   Technic Shock Absorber 9.5L (75348). This type of piece allows for a LEGO creation to literally have a spring in it. Note the cross hole on one end and round hole on the other.

116.   Technic Bearing Plate 2 x 2 (2444). This is essentially a 2 x 2 LEGO Plate with a round hole on one side.

117.   Technic Double Bearing Plate 2 x 2 (2817). Although this looks similar to the Technic Bearing Plate, this is a 2 x 2 Plate with a round hole on two sides opposite from each other.

118.   Technic Shock Absorber 6.5L (76537). This is like the previously mentioned shock absorber,b uts mallera ndw ithn oc ross-holes ection.

TechnicM iscellaneousP ieces

Yes, I could explain a lot about what pieces constitute the Miscellaneous Pieces, but I will introduce many of these as I go. I have tried to build these models with the basic sort of pieces that I have explained above. Figure 1-38 presents a collection of these odd pieces.

Figure 1-38. A collection of some Miscellaneous Pieces from Technic

119.   Technic Cross Axle Extension, Ribbed (6538). This is another way to join two Axles together, and these are ribbed on the side, making them easy to pull off.

120.   Technic Tube 2M (75535). This is a way to put a tube over 2M of an Axle for increased stability, and you can also join a Connector Peg to it as well.

121.   Technic Cross Axle Extension (59443). This is another type of Cross Axle extension, but it doesn'th avea nyr ibs.

122.   Technic Catch with Cross Hole (32039). This is essentially two cross holes, put at 90 degrees to each other.

123.   Technic Catch (6553). This is essentially an Axle with a Bush placed 90 degrees of it.

124. Technic Toggle Joint (32126). This is a Bush with a ring attached to it, and two of these joined together can make the Axle join at any angle.

125. Technic Change-over Catch (6641). This is a way to perfect shifting on a Technic creation.

126. Universal Joint (61903). This allows an axle to go at any angle provided it is below 90 degrees. This is especially helpful when creating vehicles with spinning axles that need tob end.

Designing Digitally with LEGO Software

Now that you have your LEGO pieces together, it may be worth your time to check out some software made for designing with LEGO. Of course, there are some limits and non-limits for digital programs. You might be able to create a structure that looks like it would work well in real life but doesn't follow the basic laws of gravity. I have some examples in the next chapter of that.

These LEGO software programs are like a drawing board, one that you will probably have to go back to from time to time. Once you can get your LEGO creation to work in a three-dimensional (3D) program, you can then build it in real life.

Some of you might be just as comfortable, if not more comfortable, to get out your actual LEGO Technic pieces and just continue to build until you find a model that works. I generally like to build a model in real life and then work out any problems with the model in the digital version. Then I can rebuild the LEGO model in real life and do the digital version again. However, if you want to build a LEGO robot in a digital blueprint, the following sections will describe some of the software programs that I would recommend trying.

LEGOD igitalD esigner

I found that the simplest of programs for creating LEGO structures in a 3D LEGO digital space is LDD, which is created by LEGO itself. It can be downloaded on the official LEGO site at `http://ldd.LEGO.com/`. It is free for use with Windowsa ndM ac.

Using LDD is as simple as pulling a piece from the side menu, which groups them in such a way so it is simple to find the piece you are looking for. There is even a search engine that allows you to search by part number or name. Once you find the piece and color, it is as simple as dragging and placing. For example, if you have a Connector Peg and a Beam, it is simple to pull and drag the Peg to the Beam and place it in the proper place. The pieces then have a cool green outline to them, and you can sink them in place properly.

Other special features include cutting and pasting pieces or groups of pieces, which is very helpful in designing creations where the same piece is used over and over again. There is also a Hinge tool for times where you need to turn a piece on a 360-degree axis, a Hinge Align tool that allows you to line up LEGO pieces perfectly, a Flex Tool for flexible pieces, and a Paint Tool so you can paint pieces different colors.

One of the most amazing features is the Building Guide Mode. This comes in handy when you already have a completed model; the program will show you how to build it step by step, as if you were creating a set of LEGO building instructions like those that come with every LEGO set.

I wish I could show you a screenshot, but LEGO does not allow the program to be used for commercial purposes. But I highly recommend downloading it and using it to create your creations, as all the models that you will see in this book were originally created on LDD. To get the screenshots in the next sections, I used MLCAD and LDraw.

MLCAD

MLCAD is developed by Peeron and it can be downloaded at `http://mlcad.lm-software.com/`. M LCADi ss imilar to LDD, but it involves greater control over LEGO pieces.

You can export an LDD file to an MLCAD or LDR file through this series of steps (Figure 1-39).

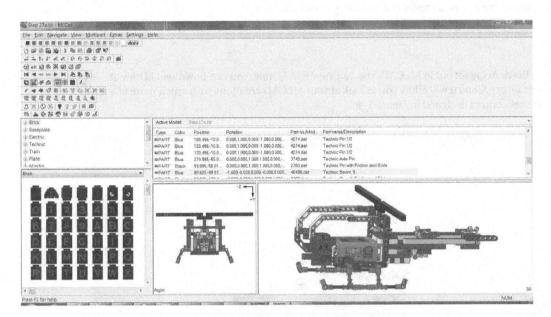

Figure 1-39. *A screenshot of a LEGO model on MLCAD, a model that was originally designed in LDD*

1. In LDD, open the model you wish to export. Click File on the Navigation bar and go to Export.

2. A Save As window will open, so designate where you want to save your model and give it a name.

3. Under Save As Type select LDraw Files.

4. ClickS ave.

5. Openy ourf ilei nM LCAD.

I have found that when I have exported certain models from LDD to MLCAD, some pieces do not show up. I then have to put in the pieces manually on MLCAD. You can see the menu of parts on the bottom left corner in Figure 1-39, and there is a parts browser with the part number and name. Keep in mind that what LDD and MLCAD call an individual LEGO part is not necessarily the same name (the pieces mentioned by name earlier in this chapter were the LDD names).

In addition to the parts window, there is a viewing window that allows you to select individual pieces to move them, change their color, and rotate them. There is a chart directly above it that will have individual information like type, color, position, rotation, part number/model, and part name/description. To the right of the viewing window, there is a display where you can turn the model around in three dimensions, so you can really see what you are creating.

One of the great advantages of MLCAD is that you are not limited by the amount of colors. LDD has certain pieces that are not available in certain colors, but MLCAD allows you to change a color by selecting a part in the viewing window and then selecting the proper color.

I found the most useful tools are the rotation and moving tools. This allows you to rotate a piece or move it into its proper place. You will find that MLCAD does not limit where you want to put pieces, but LDD will literally not allow you to put two pieces together that won't fit together. If you want the same piece to occupy the same space as another piece, you can do it. Granted, the final model will not be physically possible to build, unless you can fuse LEGO pieces together on an atomic level.

When it comes time to view a model realistically, you can use LDraw.

LDraw

Once you have the drawing set out in MLCAD, you can view it in LDraw. You can download LDraw at http://www.ldraw.org/, and it will allow you to look at your MLCAD creations with much more of a realistic view. You can see the helicopter I designed in Figure 1-40.

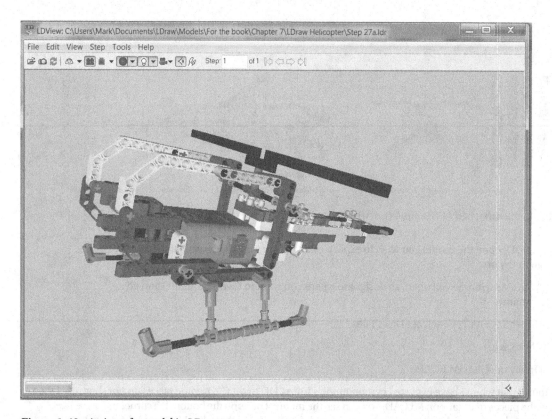

Figure 1-40. *A view of a model in LDraw*

From here, you can view your creation in a 3D view and turn it for viewing in three dimensions, almost like you are holding it in your hand. LDraw features all sorts of tools so you can change the background, take a screenshot, and others that I encourage you to try out.

Summary

For assembling a LEGO Technic robot kit, I suggest first going to the Pick a Brick section on the official LEGO web site or to BrickLink. Once you have purchased all the parts you need, you can go to Peeron, Brickfactory, or the official LEGO web site to get instructions for LEGO models.

For your LEGO Technic robot, you will need a healthy supply of Bricks (both traditional and Technic), Beams, Levers, Axles, Connector Pegs, Bushes, Cross Blocks, Angle Elements, Gears, Racks, Shock Absorbers, and various other Technic Miscellaneous Pieces. I highly suggest purchasing ample amounts of all pieces described in this chapter and storing them in a container like a tackle box so they are organized properly.

If you are looking to start building but want to try it in a digital world, there are many programs for that. LEGO provides a free program on its web site known as LEGO Digital Designer (LDD), and there are also MLCAD and LDraw. These programs allow you to build a model piece by piece and view it in a 3D world. They are at least worth looking into.

The next chapter will go into detail about how to uses these parts to fit together to form all kinds of shapes for ar obotb ody.

CHAPTER 2

■ ■ ■

Creating a Robot Body

One of the reasons I like to build robots with LEGO Technic is that I have always admired robots from speculative fiction films. Remember R2-D2, or C-3PO, or WALL-E, or The Iron Giant? All these fictional robots have one thing in common: they have a body. Sometimes the body was shaped almost exactly like a human, and sometimes the shape was something not drawn from nature at all.

In this chapter I am going to discuss how to use LEGO Technic bricks to create the basic shape of a body for your robot and how to employ my three laws of LEGO Technic robotics in doing so. I am also going to discuss how to create wireframe models, which is something I discussed in my last LEGO Technic book as well. Although I will show you how to make shapes such as squares, rectangles, triangles, and trapezoids, how you build your robot body is entirely up to you.

This will serve as a basis for the next chapter, where you'll learn how to create a robot base with wheels. So let's dig in by first reviewing my three golden rules of LEGO Technic robotics.

The Three Laws of LEGO Technic Robotics

It is my belief that the short stories of Isaac Asimov were instrumental in our modern understanding of robotics. In fact, the word robotics was even derived from one of his short stories "Runaround," which you can read in the book *I, Robot*, a collection of robot short stories. *I, Robot* the book is very different from the 2004 Will Smith movie of the same name, but both the film and the original book feature the three laws of robotics, which are a foundation for Asimov's fictional robots. His three laws of robotics are essentially as follows:

1. Robots should not hurt humans or allow humans to get hurt.

2. Robots should abide by the rules given to them by humans, except when the rules conflict with law #1.

3. Robots should protect themselves, except when such protection conflicts with law #1 or law# 2.

Although you will not have to deal with Isaac Asimov's three laws as you create your LEGO Technic automatons, you will have to follow other laws when you are creating LEGO Technic robots. As Asimov has his three laws of robotics in his works of robotic fiction, I also have my three laws of LEGO Technic robotics. We will adhere to these throughout thec hapter:

1. A robot needs room for its inner mechanizations, and it is best to plan before one builds.

2. A robot must be built in a way that allows its pieces to be used in a proper fashion.

3. A robot must follow the laws of geometry (demonstrated in Projects 2-1, 2-2, and 2-3).

The First Law of Technic Robotics: Planning

In my first book on LEGO Technic, I discussed the concept of wireframing. That book discusses how to build LEGO objects such as cars, planes, and various other vehicles. You will discover that when you build a robot, some of the work will be done on the exterior shell that surrounds the inner workings. It is important that you have a good idea of how to construct the outside shell, and that you leave some space inside it for the LEGO Power Functions, which we will discuss in the next chapter. To form an exterior shell, I like to use wireframing, which is a term I use for a simple visual presentation of a physical object in 3D computer graphics by drawing lines at the location on each edge.

If you have ever taken a drawing course, you would have learned how to break down a complex object into basic shapes like spheres, cones, cylinders, and so on. In the same manner, you can learn how to "think in LEGO" and see what an object would look like if it were made of Technic LEGO pieces.

You can see in Figure 2-1 a picture of a Smart Car on the left. Imagine looking at the Smart Car through a glass window and placing tracing paper on the glass. You will then see that you have something in the center. If you try to imitate the shape using the LEGO Technic pieces, you would form the crude shape you see on the right.

Figure 2-1. *A demonstration of how to create a Smart Car in wireframe, and then with LEGO Technic pieces*

When it comes to robots, there are many types that are used in the real world, but the most fun ones still only exist in speculative films and literature. If you have ever looked at how such films are made you will discover that there are several concept sketches in the planning stage when it comes to films with robots. Several concept drawings of robots usually end up getting thrown away, but for some reason, certain robots get made. Perhaps it is something in the basic shape that draws people in. As you build with LEGO Technic (or any type of LEGO), you will soon discover your own personal code of aesthetics.

You will discover that you like the look of certain shapes, and you may not even be able to explain why. For example, the robot EVE from *WALL-E* was designed by Jonathan Ive, who also helped with the design of the iMac, iPhone, and iPad. Don't you think EVE looks like something that Apple would put out if it were in the robot-making business?

However you want build, the software programs discussed in Chapter 1 (LDD, MLCAD, and LDraw) are definitely worth trying. These LEGO software programs are like a drawing board, one that you will probably go back to often. Once you can get your LEGO creation to work in a 3D program, you can then build it in real life.

The Second Law of Technic Robotics: Build It Strong

Basically, LEGO Technic creations need constant reinforcing in order to work. So make sure you build your projects strong. I will give you an example from the robot wheeled base that I discuss in the next chapter.

My first attempt at mounting an XL-Motor above the shaft was connected with two gears. I found I could make the robot base go forward, but when I attempted to go backward, the gears jammed. The reason why had to do with how I mounted the engine, on Zero Degree Elements. As the car traveled along, the weight of the motor was insufficient to keep it held down, so it constantly shifted out of place. The solution was to mount the engine below, so it was secured by the shaft. Even then, I found that the engine kept slipping, so I secured it even more.

In building a LEGO model, you have to secure pieces in place and often doubly secure them. This is the case of the M-Motors from Project 3-1. You will note that it is secured in the front in two places, so it can take control of the propulsion, and if the Connector Peg were not there, the entire spinning motion would rest in that one place. I wanted to secure it from the bottom as well as from the front, which is why I put in a Technic Brick and locked it in. It would have been better if the M-Motor had side holes, but you have to work with what you have with Technic LEGO parts.

LEGO Technic parts change constantly with every new set, and you will find that a recent set (9398) is similar to those shown in Project 3-1 through Projects 3-3 in that it has four-wheel drive, four-wheel steering, and suspension. I found that this set could not be created with the pieces I had, or the pieces I detailed in Chapter 1. Since LEGO has the power of the pieces, they will often create pieces that will help builders even more.

The Third Law of Technic Robotics: Geometry

In Projects 2-1, 2-2, and 2-3 you'll find that you can learn a lot of geometry while playing with simple LEGO pieces, and you will also discover that everything you learned about in high school geometry class applies to building with LEGO Technic.

Before you renew your geometry skills though, let's construct our LEGO base in Project 2-1.

■ **Note**   Before beginning this chapter's projects, refer to Appendix A for a complete list of required parts.

Project 2-1: Creating a LEGO Base

The LEGO base we'll build can serve as a foundation for any LEGO robot (Figures 2-2 through 2-16). In just 15 steps, you'll see how. You should recognize the pieces, as I discussed them in Chapter 1. The exception is the Battery box, which I cover in then ext chapter, B ring Y our L EGO T echnic R obots to L ife w ith P ower F unctions.

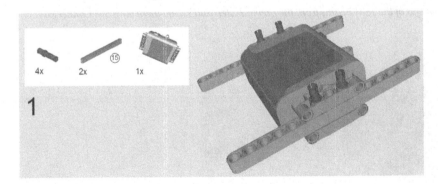

Figure 2-2. *Center a 15M Beam on each side of the Battery box and secure them with 3M Connector Pegs*

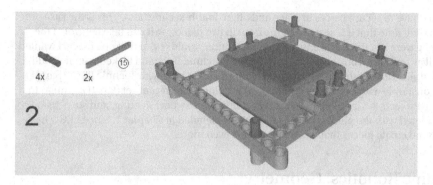

Figure 2-3. *Place two 15M Beams at 90-degree angles to the ones from the previous step and secure them with 3M Connector Pegs*

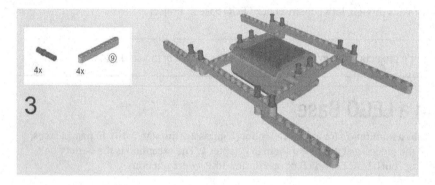

Figure 2-4. *Place a 9M Beam at each end of the 15M Beams from the previous step and secure them with 3M Connectors*

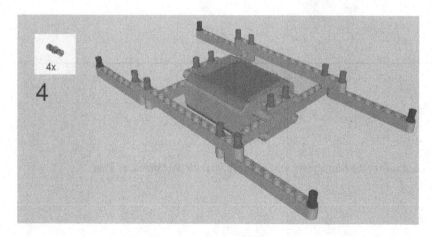

Figure 2-5. *Attach a Connector Peg to the ends of the 9M Beams and place a 5 × 3 Angular Beam atop them*

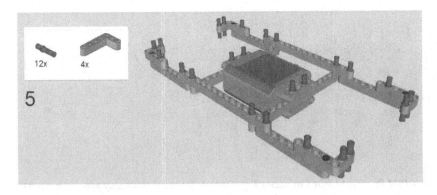

Figure 2-6. *Place three 3M Connector Pegs on each of the 5 × 3 Angular Beams*

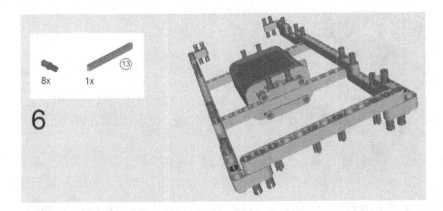

Figure 2-7. *On one side of the construction, snap a 13M Beam into place, and insert four Connector Peg/Cross Axles below with the Axle end down. Place a Connector Peg/Cross Axle on the spots as shown, with the Axle end up*

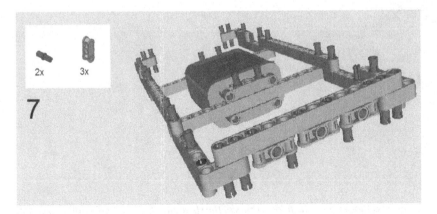

Figure 2-8. *Attach three Double Cross Blocks below. Note two of the Double Cross Blocks attach only in one place, and a Connector Peg/Cross Axle goes in this other spot*

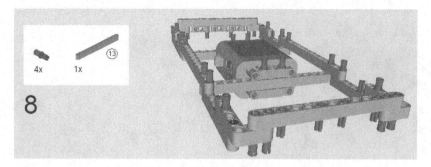

Figure 2-9. *On the other side, snap in a 13M Beam and four Connector Peg/Cross Axles in the same way you did in Step 6 (Figure 2-7)*

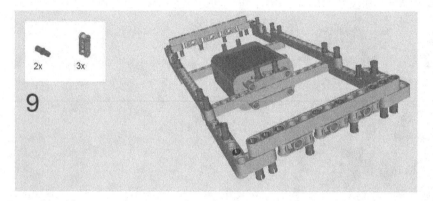

Figure 2-10. *This is another step where you need to mirror the other side. Insert the pieces as shown, like you did in Step 7 (Figure 2-8)*

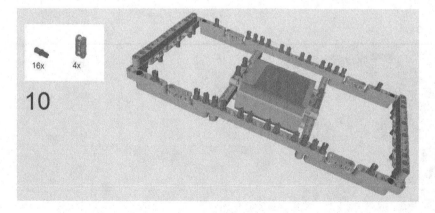

Figure 2-11. *You can see that the Double Cross Blocks go in the four corners of the structure. The 16 Connector Peg/Cross Axles go in the many holes in the center of the structure*

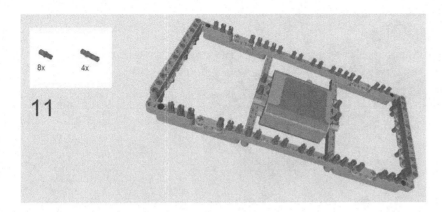

Figure 2-12. Another 3M Connector Peg goes in each corner, along with a Connector Peg/Cross Axle. Place Connector Peg/Cross Axles on top of the Double Cross Blocks

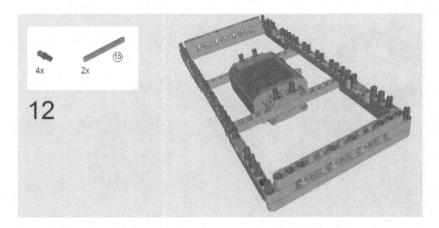

Figure 2-13. Each of the corners gets a Connector Peg/Cross Axle, and two 15M Beams go on each side

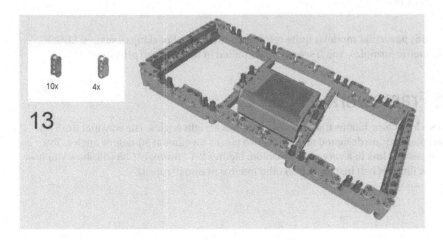

Figure 2-14. Five Double Cross Blocks go on each side, atop most of the Connector Peg/Cross Axles. Note the placement of the 3M Cross Blocks

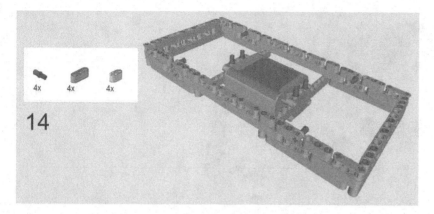

Figure 2-15. *Place the 3M Beams on each corner, along with the 2M Cross and Hole Beams. The Connector Peg/Cross Axles go in the 3M Cross Blocks from the previous step*

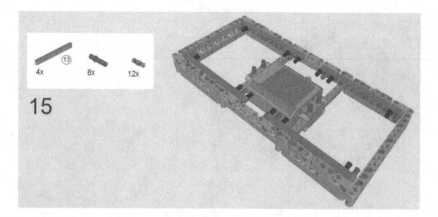

Figure 2-16. *Insert the 3M Connector Pegs in the sides of the Double Cross Blocks in each corner. Snap on the 13M Beams on each corner, and then insert the Connector Pegs*

You will notice that the base of this particular model is quite rectangular in shape. Speaking of shapes, LEGO models require an allegiance to geometric formulas. You'll see this exemplified in the next two projects.

Project 2-2: A Solid Framework

You will notice that Project 2-2 uses a lot of long beams that meet one another at right angles. The way that the 5×3 Angular Beams meet is a good example; they are designed to connect two pieces together at 90-degree angles. You will find that pieces linked at 90 degrees will link in a very strong fashion. Figures 2-17 through 2-38 will show you how tob uilde venh igheri n2 2s teps.Y ouw illn otet hatt hisp rojecti sb uiltd irectlyo nt opo fP roject2 -1.

Figure 2-17. *Connect the 4 × 2 Beams on each side of the structure and insert a Connector Peg and Connector Peg/Cross Axle in there*

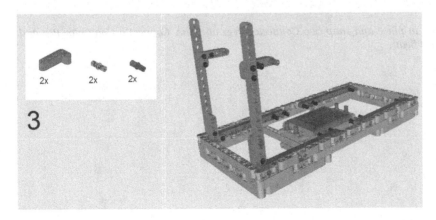

Figure 2-18. *Snap on a 15M Beam on each side and then insert two Connector Pegs on each side*

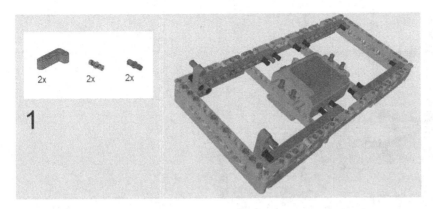

Figure 2-19. *Snap on two 4 × 2 Angular Beams and insert a Connector Peg and a Connector Peg/Cross Axle on each of the Angular Beams*

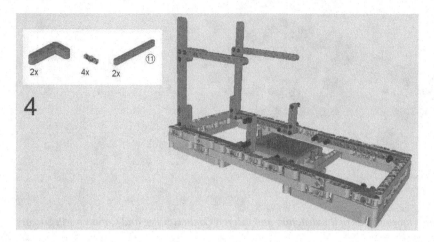

Figure 2-20. *Connect 11M Beams on the 4×2 Angular Beams from the previous step. Snap on the 5×3 Angular Beams and place two Connector Pegs on each of them*

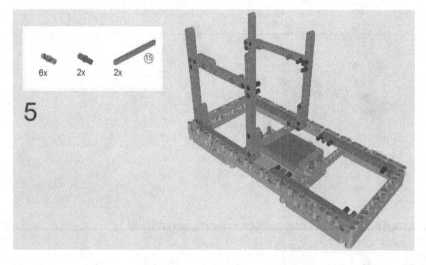

Figure 2-21. *Snap two 15M Beams in place and snap two Connector Pegs on these. Connect a Connector Peg and Connector Peg/Cross Axle on the 11M Beam*

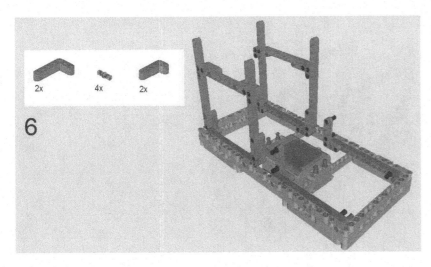

Figure 2-22. *Two 4×2 Angular Beams snap into place up high on the creation, while two 5×3 Angular Beams attach below. Don't forget to put the Connector Pegs in place on the 5×3 Beams*

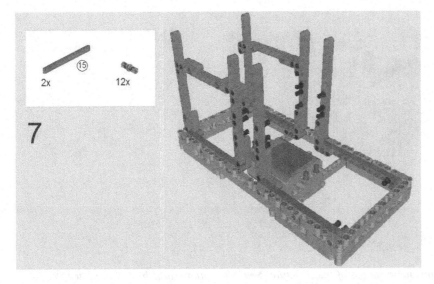

Figure 2-23. *Snap two 15M Beams into place and place twelve Connector Pegs into place on those and the other 15M Beams*

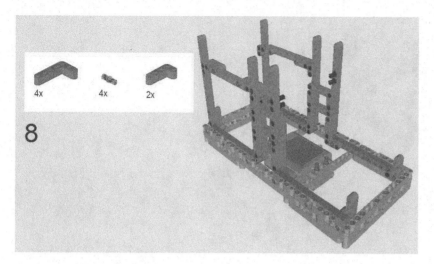

Figure 2-24. *Snap the four 5×3 Angular Beams in place. Place the Connector Pegs above them. Near the bottom of the creation, attach the 4×2 Angular Beams*

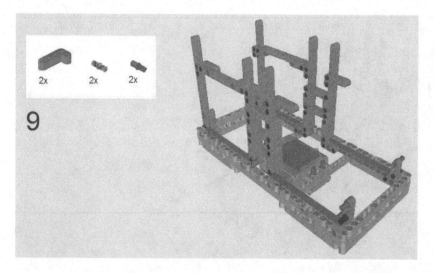

Figure 2-25. *Attach two 4×2 Beams near the top of the structure. Near the bottom, attach the Connector Pegs and Connector Peg/Cross Axles*

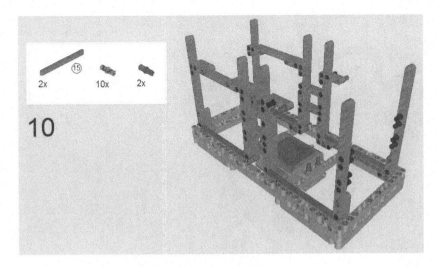

Figure 2-26. *Snap in 15M Beams on the 4 × 2 Angular Beams. Attach eight Connector Pegs on the 15M Beams. Attach two Connector Pegs and Connector Peg/Cross Axles on the 4 × 2 Beams on the top*

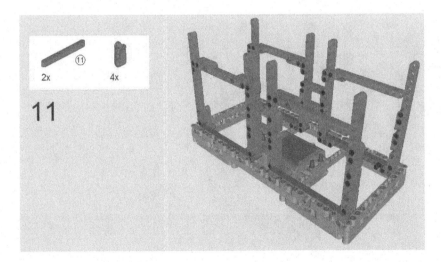

Figure 2-27. *Snap on an 11M Beam on each 4 × 2 Angular Beam above. Snap in the four 3M Cross Blocks*

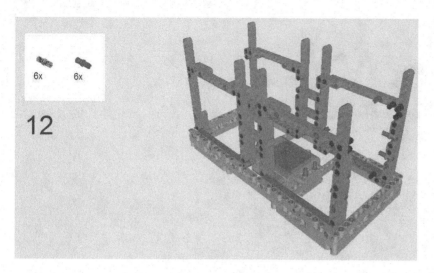

Figure 2-28. *Attach the six Connector Pegs in the corners as shown. Four of the Connector Peg/Cross Axles go on the 3M Cross Blocks and two more go on the 11M Beams*

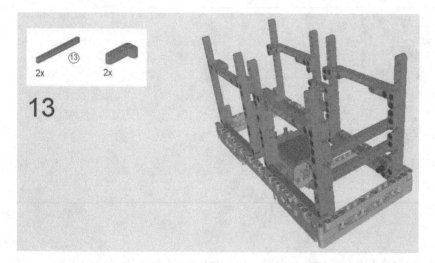

Figure 2-29. *Snap on the 4 × 2 Angular Beams. The 13M Beams connect together the left and right sides of the structure*

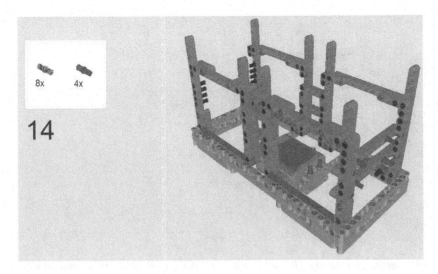

Figure 2-30. *The four Connector Peg/Cross Axles go on the 13M Beams. The eight Connector Pegs go on the 15M Beams on the opposite side*

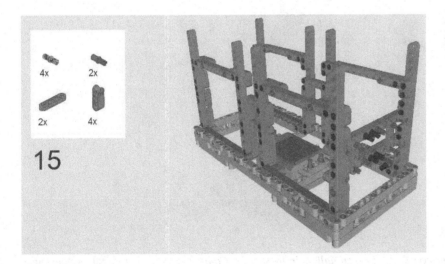

Figure 2-31. *The 3M Cross Blocks go on the eight Connector Pegs from the previous step. Four Connector Pegs and two Connector Peg/Cross Axles go on the 13M Beams. Two 4M Levers help bridge the 13M Beams*

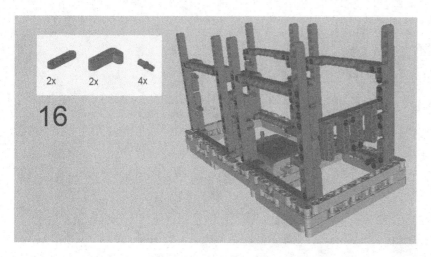

Figure 2-32. *Another two 4M Levers go atop the other Levers. The two 4 × 2 Angular Beams join here, which bond the side very well. Add the four Connector Peg/Cross Axles on the 3M Cross Blocks on the opposite side of the creation*

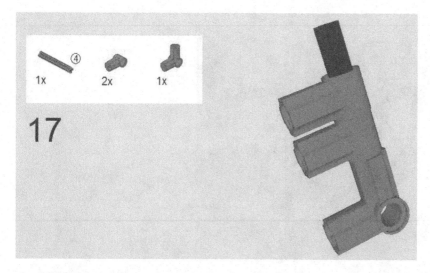

Figure 2-33. *This is a separate project that will eventually join in Step 22. Connect a 4M Axle to a 90 Degree Angle Element, and two Zero Degree Elements*

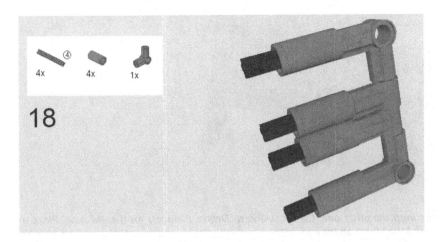

Figure 2-34. Another 90 Degree Element joins on this construction. Insert 4M Axles on each protruding cross-shaped end and place a 2M Tube over these

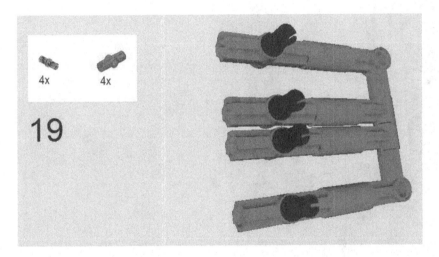

Figure 2-35. Insert a 180 Degree Element on each end of the 4M Axles. Place a Connector Peg through each round hole

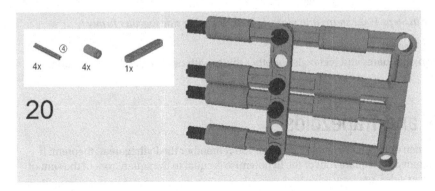

Figure 2-36. A 6M Lever joins all the 180 Degree Pieces together well. Insert four 4M Axles and four 2M Tubes

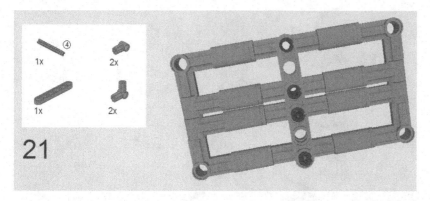

Figure 2-37. *Another 6M Lever goes atop the other one. Center two Zero Degree Elements on the 4M Axle. Place a 90 Degree Element on the other ends and stick them on the end*

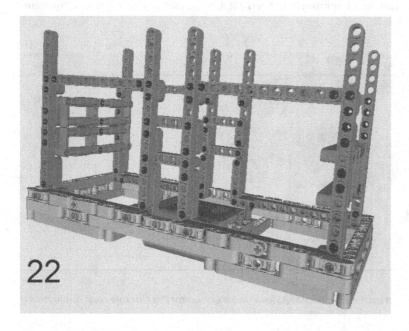

Figure 2-38. *The construction made in Steps 17-22 snaps into place, holding the other side together firmly*

Okay, you just saw many ways to do squares and rectangles or other things with right angles, but what about triangles? Project 2-3 explorest hat.

Project 2-3: Triangle and Trapezoids

If there is one thing you learn in geometry class, it is triangles. I'm sure you remember the Pythagorean theorem. If you don't, perhaps this jogs your memory: the square root of the hypotenuse is equal to the square root of the sum of the squares of its opposite and adjacent sides, or $c^2 = a^2 + b^2$.

I remember that when I was a kid, I tried to create a right triangle out of LEGO plates by using a 3M piece at a right angle with a 4M piece. Then I tried linking the other ends of those pieces with a 5M piece as a hypotenuse. Normally, a right triangle with opposite and adjacent sides of 3 and 4 will have a hypotenuse of 5. This does not work in LEGO, simply because the studs don't line up properly.

However, it should be noted that several of the Angular Beams discussed in Chapter 1, which include the 3 × 7, the 4 × 6, and the 4 × 4, are all at a 53.1-degree angle. Why is this important? Because this is the same angle that a right triangle with measurements of 3, 4, and 5 would have.

The 3, 4, 5 is known as a Pythagorean triplet, because they are whole numbers that work perfectly in the Pythagorean theorem. I have found that using LEGO Technic Beams initiates a different kind of Pythagorean theorem. You will discover that 4, 5, 6 are the perfect LEGO Technic Pythagorean triplet.

In the same manner that you can create a LEGO triangle with Angular Beams, you can also create an isosceles trapezoid. For those of you who need a review of geometry, a trapezoid is a four-sided figure with one pair of parallel sides. An isosceles trapezoid has two congruent sides. You can also find a few of these in Project 2-3.

Also in 17 steps of Project 2-3 we use a few Double Angular Beams. This piece, seen in Chapter 1, has two 45-degree angles that create a perfect 90-degree angle. You can also see how it works in Figures 2-39 through 2-55.

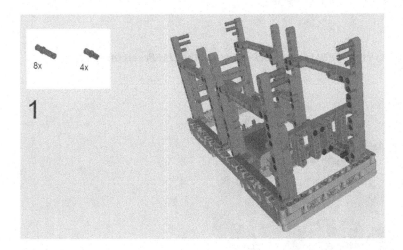

Figure 2-39. *Place eight 3M Connector Pegs and four Connector Peg/Cross Axles on top of the structure*

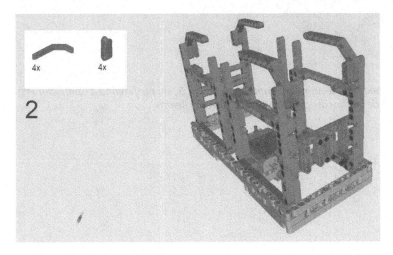

Figure 2-40. *Use a Double Cross Beam and a 3M Cross Block here on the 3M Connector Pegs and Connector Peg/Cross Axles*

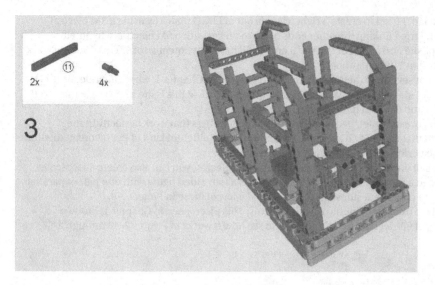

Figure 2-41. *Place the Connector Peg/Cross Axles on the 3M Cross Blocks and snap the 11M Beams into place*

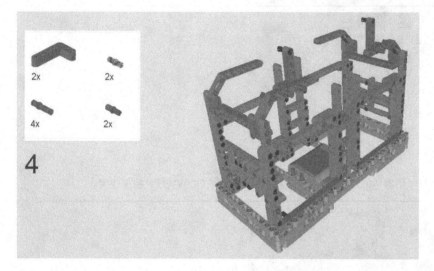

Figure 2-42. *Connect 3M Connector Pegs on two Double Angular Beams. Place a 5 × 3 Beam on that and a Connector Peg on top of that. Place a Connector Peg/Cross Axle on the corner of the Double Angular Beams*

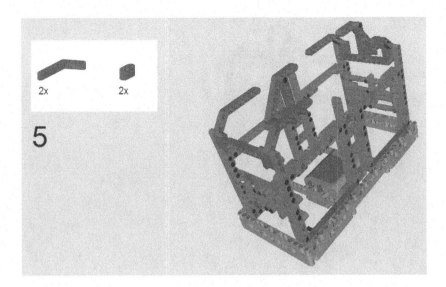

Figure 2-43. *Place a 2M Cross and Hole Beam on the Connector Peg/Cross Axle. Connect the 4 × 6 Angle Beams at the corner on the Connector Peg on top*

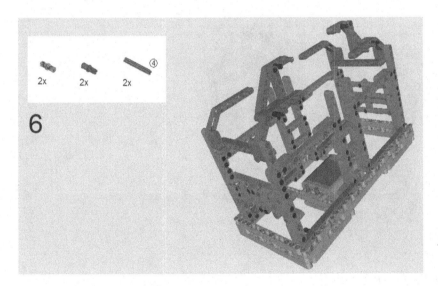

Figure 2-44. *Insert the Connector Pegs and Connector Peg/Cross Axles on the Double Angular Beams. Slide in the 4M Axle so it connects with the 4 × 6 Angular Beam*

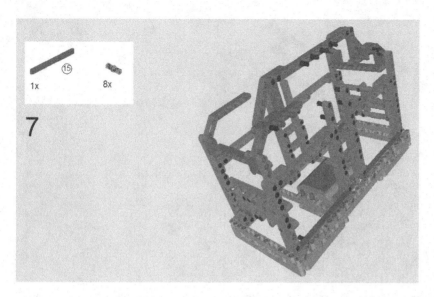

Figure 2-45. *Place a 15M Beam on top and then place four Connector Pegs there. Place two Connector Pegs on each of the 15M Beams below*

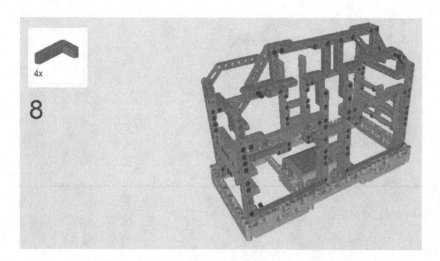

Figure 2-46. *Connect 5 × 3 Angular Beams to the structure*

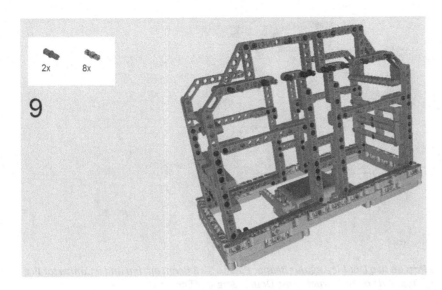

Figure 2-47. *Insert the eight Connector Pegs and two Connector Peg/Cross Axles on the top here*

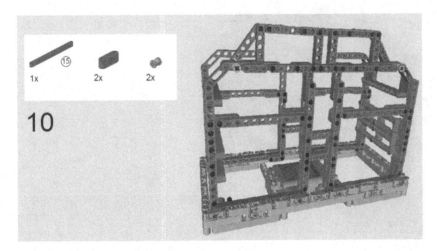

Figure 2-48. *Snap on the 15M Beam, along with the two 3M Beams. Slide on the two Bushes at the end of the 4M Axles*

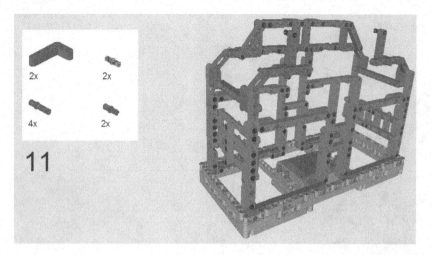

Figure 2-49. *Connect 3M Connector Pegs on two Double Angular Beams. Place a 5 × 3 Beam on that and a Connector Peg on top of that. Place a Connector Peg/Cross Axle on the corner of the Double Angular Beams*

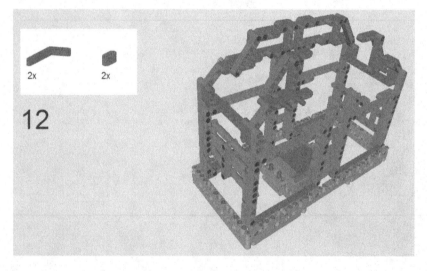

Figure 2-50. *Place a 2M Cross and Hole Beam on the Connector Peg/Cross Axle. Conenct the 4 × 6 Angle Beams at the corner on the Connector Peg on top*

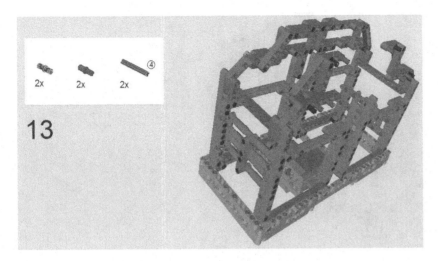

Figure 2-51. *Insert the Connector Pegs and Connector Peg/Cross Axles on the Double Angular Beams. Slide in the 4M Axle so it connects with the 4 × 6 Angular Beam*

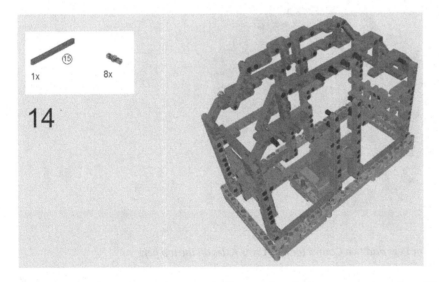

Figure 2-52. *Place a 15M Beam on top and then place four Connector Pegs there. Place two Connector Pegs on each of the 15M Beams below*

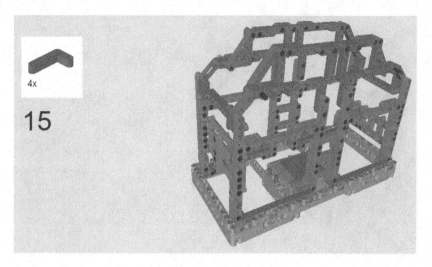

Figure 2-53. Connect 5 × 3 Angular Beams to the structure

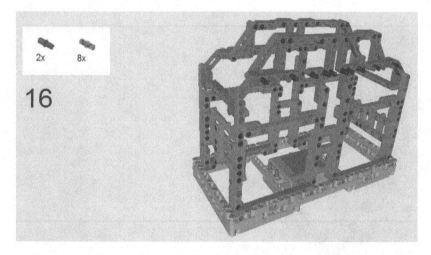

Figure 2-54. Insert the eight Connector Pegs and two Connector Peg/Cross Axles on the top here

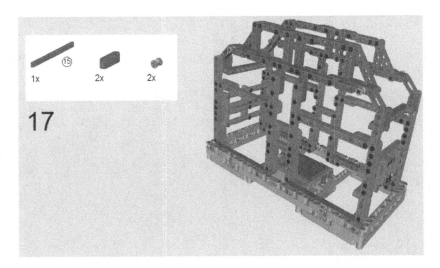

Figure 2-55. *Snap on the 15M Beam, along with the two 3M Beams. Slide on the two Bushes at the end of the 4M Axles*

Summary

Isaac Asimov uses his three laws of robotics in his works of robotic fiction, and I have my three laws of LEGO Technic robotics. In this chapter we applied them all in building three projects.

The first law is that a robot needs room for its inner mechanizations, and it is best to plan for that before you build. Fortunately, there are many software programs you can use to build a LEGO creation in a 3D space, as discussed in the previous chapter.

The second law is that a robot must be built in a way that allows its pieces to be used in a proper fashion. This means that a Technic robot creation has to be built in a way so that it is robust and will stand up to the regular wear and tear of life.

The third law is that a robot must follow the laws of geometry. Although LEGO Technic pieces are best for squares and rectangles, they can also be used to form triangles and trapezoids.

The next chapter will show you how to create a robot base with wheels which you can take control of with the LEGOP owerF unctions.

■ ■ ■

CHAPTER 3

Bring Your LEGO Technic Robots to Life with Power Functions

I mentioned in the book's Introduction that Mary Shelley's *Frankenstein* was, at its heart, a robot story. I also mentioned how Frankenstein was not the monster, but the name of the doctor who created it. Hollywood has changed the story from its original 1818 story, but there is one thing that we all remember from the 1931 movie. It is the scene where Dr. Frankenstein, after creating his monster, excitedly yells "It's Alive! It's Alive!"

In this chapter's three projects, I'm going to show you how to use Power Functions to bring your LEGO Technic robots to life. Well, they will move so they will look alive. As I said in Chapter 2, this is not a book on how to make sentient robots. Then I'm going to discuss how to create a wheeled base for a robot, which is essentially a review of some of the chapters from my first LEGO Technic book.

But first, let's review the LEGO Power Functions you'll be using.

LEGO Power Functions

I spent a lot of time in Chapter 1 discussing LEGO Technic pieces. I left some big ones out, as I believe they are important enough to have their own chapter: the Power Functions. LEGO has been in the motorized brick business since the 1970s, and some of the earliest models imitate the same design of the modern Power Functions. LEGO Power Functions come in three basic types:

- *Power*: The energy source to get a LEGO Power Function running

- *Action*: The specific purpose of what the LEGO Power Function is supposed to do

- *Control*: The method of remotely operating a LEGO Power Function

You can purchase LEGO Power Functions in the same place where you can buy LEGO Technic pieces, as detailed in Chapter 1.

Power

The power refers to the battery pack or any other method to empower any LEGO motor or similar device. There are essentially three battery packs that are available from the LEGO web site. In this book, I am going to use one power source with this battery box with six AA batteries, the 8881 (Figure 3-1).

Figure 3-1. *The LEGO Power Functions 8881 Battery Box*

The 8881 Battery Box is good for 1.5 V of power, and it can power two XL-Motors or four M-Motors, which I will explain later in the "Action" section. The Power Functions plug is on the top right next to the switch, and the switch has three positions: one for forward, the middle for off, and the other for reverse. In between the Power Functions plug is a green LED light to indicate it is working.

Most of the creations in this book will focus on the 8881 Battery Pack, as it is one of the most effective ways of powering a LEGO Technic creation. It is also relatively light and can be easily mounted on LEGO creations. I highly recommend getting rechargeable AA batteries for this, as you will go through a lot of batteries when you bring your creations to life.

Batteries are not the only way to power your creations. The 8878, as seen in Figure 3-2, has built-in lithium polymer batteries. It needs to be charged with the 8887 Transformer 10 V DC plug-in cable, which has to be purchased separately. I didn't really use it in this book, because most of the constructions are "studless" and the 8878 can be mounted easier on creations that use traditional LEGO bricks. The 8878 piece is lighter than a typical battery box, and it has seven speeds that are adjustable by turning left and right on the orange dial.

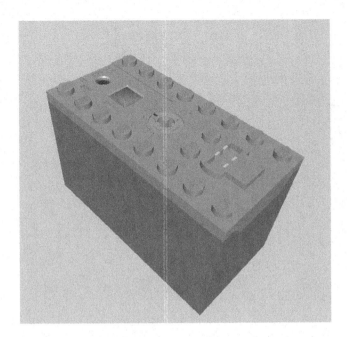

Figure 3-2. *Another method of powering your LEGO Technic creations, the 8878*

The 8881 and the 8878 are not the only ways to power your LEGO creations. There is also the 88000 AAA Battery Box, which takes six AAA batteries instead of AAs. Like the 8881, it can handle four connections on the LEGO Power Functions system with two for power supply and two for control. There are lots of studs and it is the size of 4 × 8 ×4 m odules.

Action

Once you have the power source for your LEGO Technic creations, the only thing left to do is to attach an action element to it. There are two types of action elements you can use: motors and lights.

Motors

The LEGO Technic Power Functions motors have a place where you can stick in an axle, and it will turn. The challenge is to use that spinning axle for turning the wheels of a creation to make it go. They also come in handy when taking remote control of a steering mechanism of a vehicle.

I will start with the 8882 XL-Motor (Figure 3-3). It is a super strong motor that LEGO recommends "to animate larger builds." I usually use this to power wheels, and I like it because it has two round holes on each side to mount it. It also has six round holes in front for mounting. Anything that can make a part connect in more places is better, especially for more robust studless LEGO creations.

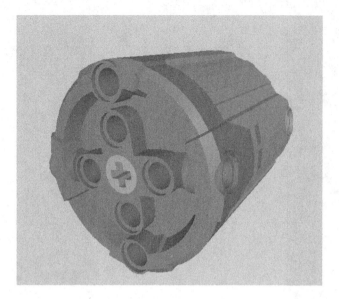

Figure 3-3. *The 8882 XL-Motor*

If you are looking for a motor with a little less of a kick, you might want to consider the M-Motor. The M-Motor differs because it creates some spin, but it does not have any round holes on the side for studless connectivity (Figure 3-4). It also has four holes on the front for holding, and the entire bottom is 2 × 6M, which is good for connectingtostudde dpie ces.

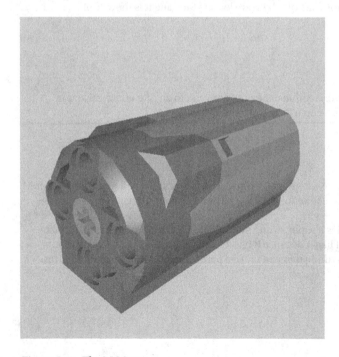

Figure 3-4. *The M-Motor*

You will notice that it is very simple to connect a motor to the battery box via the ports, as can be seeni nF igure3-5 .

Figure 3-5. *How to connect the motor to the battery box for motion*

If you turn on the battery box in Figure 3-5, the orange portion within will spin. This will turn whatever it is connected to, which in this case is an Axle and a Gear. Now this "starts the ball rolling" when it comes to bringing LEGO creations to life.

It is easy to connect two motors to one battery box. When the battery box in Figure 3-6 is powered on, the motors will spin in the same direction unless the switch is flipped to the other setting, where both motors will spin in the opposite direction. I will discuss how to properly control multiple motors in the next section box.

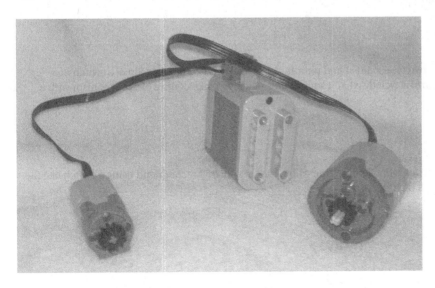

Figure 3-6. *Connecting two engines to the battery box*

Lights

Right now, while I am on the subject of power and action, I want to discuss how to put some lights on your creation. I covered this in my earlier book on LEGO Technic, but here I will mention ways to light up your creation. You can use kit 8870 to light your creation, provided it is hooked up to a battery box. The Power Functions Light LEDs (Figure 3-7) simply fit through any round hole box.

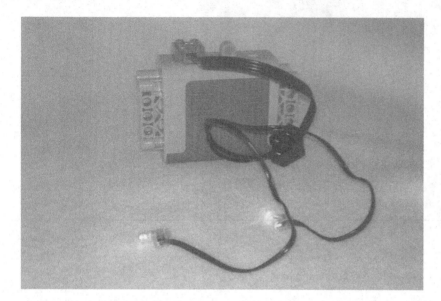

Figure 3-7. *Hooking up the Power Functions Light LEDs to the battery box*

The first thing you will learn about working with lights is that you do not want them on the entire time, as it will completely drain the battery, not to mention wear down the LEDs. I will discuss how to prevent battery and bulb burnout in the next section on "Control."

Control

When it comes to control, there are two types: direct and remote. *Direct controls* are a simple flip of a switch in a binary on/off situation. *Remote control* is exactly what it sounds like, when you take control of a mechanism with a controller.

Direct Control

Of course you aren't going to want to have the lights on the whole time, so I highly recommend using a switch like the 8869 Power Functions Control Switch (Figure 3-8).

Figure 3-8. *The 8869 Power Controls Switch*

The purpose of the power switch is to kill the battery power with a quick switch off. It works for situations where the lights need to be shut down. It can be attached as shown in Figure 3-9. You will find the switch is only off when the lever is centered. If it is all the way to the left or right, it closes the circuit and turns off the light.

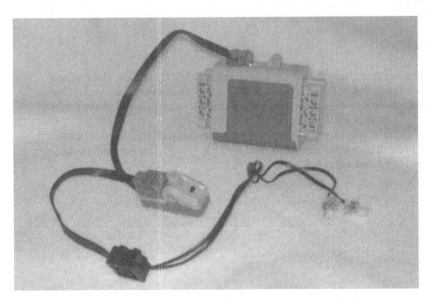

Figure 3-9. *How to connect lights with a switch. The indicator light on the battery box is on, but the LED lights are off. If the switch were flipped up or down, the LEDs would be glowing*

Remote Control

Of course you don't really want to just have constant on or constant off controls for your LEGO Technic robot. What you really want is something that will go forward and stop with the operation of a lever. In short, you want remote control, and you will need this particular IR-RX box, as shown in Figure 3-10.

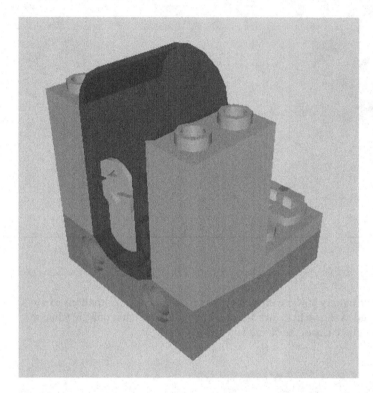

Figure 3-10. *The IR-RX, made to take control*

This is the Power Functions IR receiver that allows the user to take control of LEGO creations. This has two outputs that allow for control for two different Power Functions. You will note that it has a switch for four different channels. This is helpful for projects where many different controls are at work, but you will need to have remote controllers to back them up.

I would not recommend using a remote control for lights, as this is a constant on and off function. If you have a creation, specifically one with more than one motor, this works quite well. Simply connect them as shown in Figure 3-11, with the other end connected to the battery box.

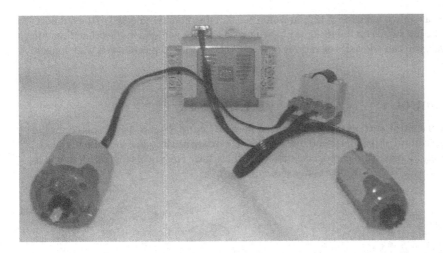

Figure 3-11. *Connecting two engines to the IR-RX*

Project 3-1 is a construction where you'll need to use the IR-RX. You will also need one of two remotes to work with it.

The IR-TX Remote

You can use the IR-TX to start, stop, and change direction (Figure 3-12). The red switch controls the red outlet of the IR-IX receiver and the blue switch controls the blue outlet. It has four channels, and it has a reaching distance of at least 30 feet (10 meters). It requires three AAA batteries to function.

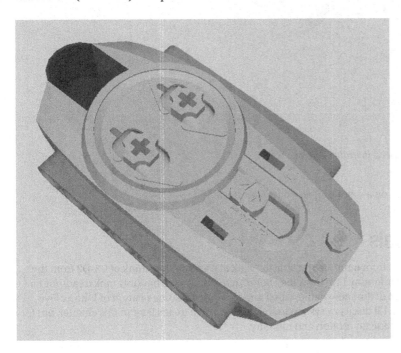

Figure 3-12. *The IR-TX Remote*

If you were to use this remote, as shown in Figure 3-12, you would see that the motor connected to the red port will rotate in one direction when the red switch on the IR-RX is pushed up. The motor will spin in the other direction when it is pushed down. By the way, you can add axles or various other cross hole–shaped pieces if you want longer lever-likec ontrols.

The IR Speed Remote Control

You can also use some other kind of control, such as the 8879 IR Speed Remote Control (Figure 3-13). This controller has knobs that you can turn to take control of speed. I found that a small flick would cause a slow increase in speed, and continuing to turn the knob only increases the speed. Like the IR-RX, these have cross holes that are designed to fit all kinds of LEGO Technic pieces, which can provide a better sense of control.

Figure 3-13. *The 8879 IR LEGO Technic Speed Remote Control*

Okay, I'm sure you are anxious to start on a robot project, so let's get started.

A Robot Base with Wheels

I think we all know that some of best robots from films are not able to "walk like a man." Just think of R2-D2 from the *Star Wars* films, with his constant rolling, or Johnny Five from the *Short Circuit* films, who has only tank treads for feet. This is also similar to WALL-E from the film of the same name, who I am convinced is some relation of Johnny Five (you have to admit there is a resemblance). I'll discuss a system for taking control of treads later in this chapter, but for now, I'll mention how to create a wheeled base for motion and mobility.

This project is going to focus on wheeled vehicles, with a lot of examples mentioned from my previous book. Here you will learn how to put a motor on wheels and how to steer it, and, as an added bonus, I'm going to endow the creation with four-wheel drive. I decided to put an even greater twist on this robot base and give it four-wheeled steering for greater control. Some of you might even remember the chapter from my previous book that dealt with shock absorbers, and I am going to use this as well.

Using an IR-RX and one of the remote controls listed above, you can make this base roam across your floor. It can serve as a foundation for your robots until you learn to make them walk, which we will discuss in Chapter 7.

Project 3-1: Creating a Four-Wheel Drive Engine Wheel Base

I wrote about how to use four-wheel drive in my first book, and for that, I managed to spin four wheels with just one Technic Motor. Just to let you know, it is possible to do it with one motor, provided the motor is located close to the axles, but this involves many gears. It worked, but the more gears you add to a project, the more likely it is to jam up later. I soon discovered that two motors can do the job with less jamming, so that's what we'll do here in Figures 3-14 through 3-34.

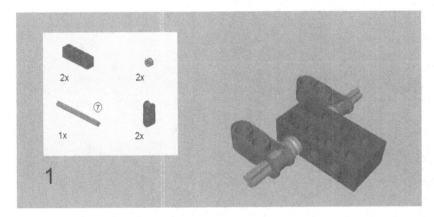

Figure 3-14. *Slide a 7M Axle through the 1 × 4 Technic Bricks and center the Axle between the two bricks. Cap them off with a Half Bush and the 3M Cross Blocks. Leave 1M of Axle on each side*

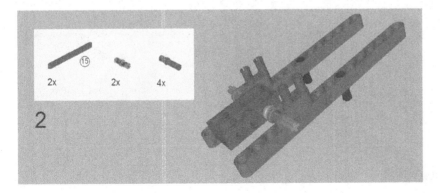

Figure 3-15. *Center the 3M Connector Pegs through the 3M Cross Blocks. Snap on the 15M Beams with and then add the Connector Pegs*

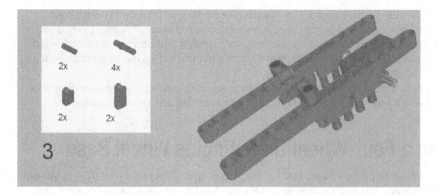

Figure 3-16. *Attach another set of 3M Connector Pegs and 3M Cross Blocks to the 15M Beams. Snap on the 90 Degree Cross Blocks and insert the 2M Axles*

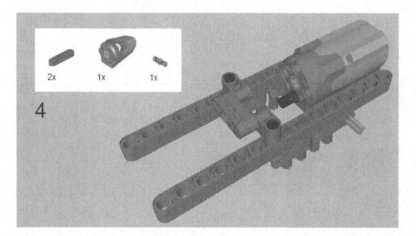

Figure 3-17. *Attach the M-Motor with the Connector Peg on the two 1 × 4 Technic bricks. Insert the two 3M Levers on the 2M Axles*

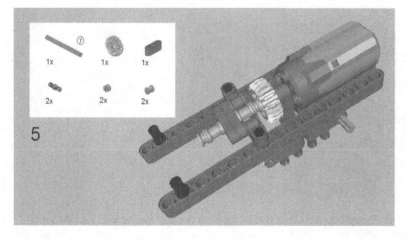

Figure 3-18. *Snap on the 3M Beam on the M-Motor and Connector Peg. Slide in the 7M Axle and make certain you have the Half Bushes, Bushes, and the Z20 Gear. Don't forget to add on the Connector Pegs, as shown*

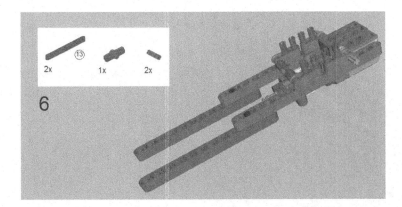

Figure 3-19. *Snap on the 13M Beams. Place the 180 Degree Angle Element in the center of the 90 Degree Cross Blocks and anchor it in place with the 2M Axles*

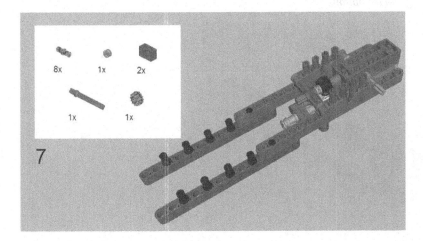

Figure 3-20. *Place the Connector Pegs on the 13M Beams. One 1 × 2 Brick goes on each end of the M-Motor. Insert the 5.5 Axle with end stop through the 180 Degree Angle Element until it goes into the 1 × 2 Technic Brick, and place the Half Bush, the Z12 Gear, and 1 × 2 Brick in between*

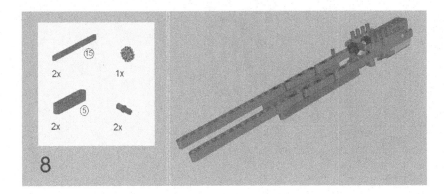

Figure 3-21. *Snap on the 15M and 5M Beams on the 13M Beams. Snap on the Connector Peg/Cross Axles, as shown. Slide a Z12 Gear on the 5.5M Axle*

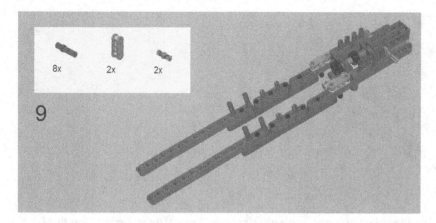

Figure 3-22. *Place the Double Cross Blocks over the Connector Peg/Cross Axles from Step 8. Place in the 3M Connector Pegs, as shown, and the Connector Pegs go on the bottom*

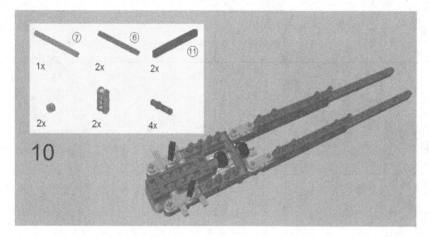

Figure 3-23. *Place the 11M Beams on top like you see here. On the ends, slide the 6M Axles vertically through the creation and place Double Cross Blocks on top of those. Slide a 7M Axle through the round hole of the Double Cross Blocks, and secure them with Half Bushes*

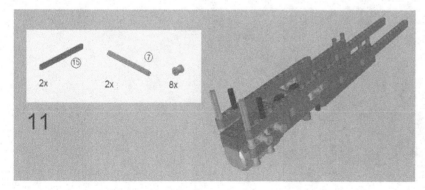

Figure 3-24. *Place a 15M Beam on top and slide on a 7M Axle. Note the placement of eight Bushes (four Bushes on each side), designed to anchor things in*

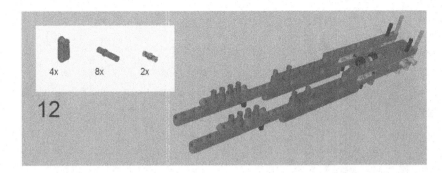

Figure 3-25. *Place the 3M Connector Pegs with 3M Cross Blocks on top and the Connector Pegs on the bottom*

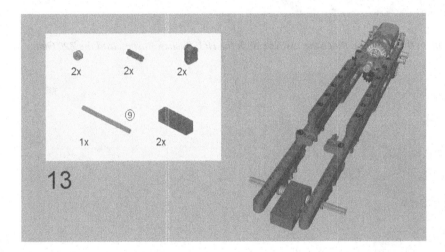

Figure 3-26. *I flipped the model so you can see what to do here. This is essentially a repeat of what is on the other side with the 9M Axle, 1 × 4 Technic Bricks, 90 Degree Angle Elements, 2M Axles, and Half Bushes*

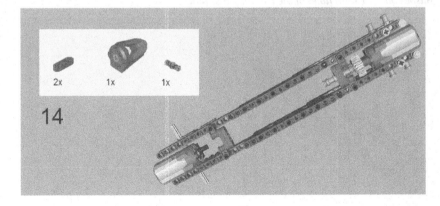

Figure 3-27. *This step is identical to Step 4 but with the M-Motor, Connector Peg, and 3M Levers on the opposite side*

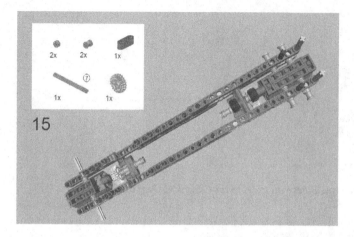

Figure 3-28. *Snap on the 3M Beam on the M-Motor. Place the 7M Axle with the Half Bushes, Bushes, and the Z20 Gear*

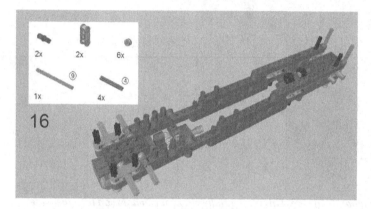

Figure 3-29. *Place the 4M Axles on the end and use the Double Cross Blocks and four Half Bushes to anchor them in place. Use the 9M Axle with Half Bushes on the round holes of the Double Cross Blocks*

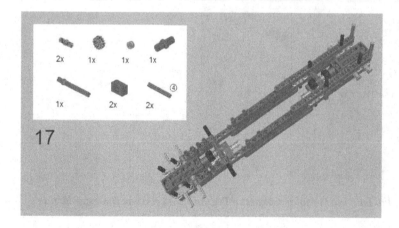

Figure 3-30. *Anchor the 180 Degree Element in place with the two 4M Beams. Place the 5.5 Axle, 1 × 2 Technic Bricks, Half Bush, and Z12 Gear so they match the other side*

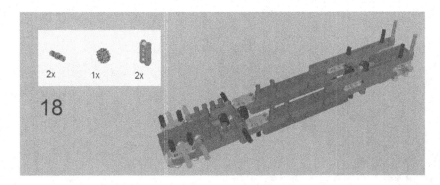

Figure 3-31. *Place the Connector Pegs on the bottom and the Double Cross Blocks on top. The Z12 Gear goes in the middle of all that*

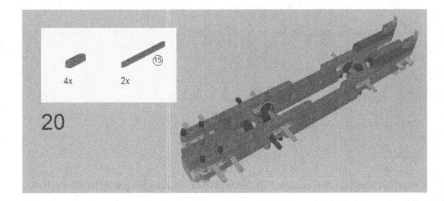

Figure 3-32. *The 9M Beams go on the end of the creation. Important to note are the 9M Axles in the middle, with each having a tube and a Z12 Gear*

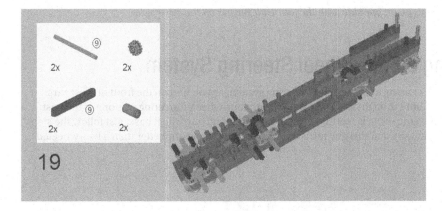

Figure 3-33. *The four 3M Levers go on one end, while the 15M Beams go on the other end*

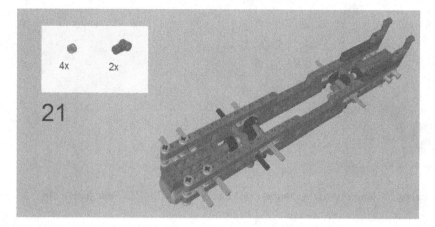

Figure 3-34. *The Zero Degree Elements go on one side with the four Half Bushes on the other side*

Project 3-2: Creating a Four-Wheel Steering System

I probably could have done this project using a two-wheeled steering system, as you find on the front of most cars. I can think of several ways to do that, but I thought that this method with four-wheeled steering was one of the best ways tod ot hat.S ot oc reatea nda dda f our-wheels teerings ystemt oo ure xistingf our-wheel base, just follow the steps in Figures 3-35 through 3-55. I wouldn't say that this guy turns on a dime, but it does turn better than a two-wheeled steerings ystem.

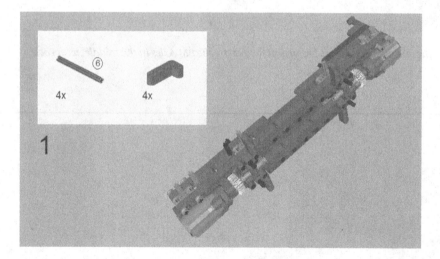

Figure 3-35. *On the underbelly of this creation, snap on the four 4 × 2 Angular Beams. Push in the 6M Axles through the round hole on the "corner" of the 4 × 2 Beams*

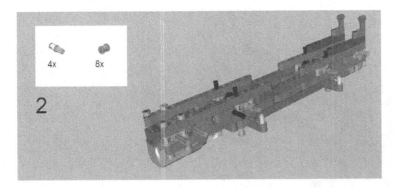

Figure 3-36. *Cap off one end of the 6M Axles from Step 1 with the Bushes. Place the other four Bushes on the two 9M Axles. Push in the Connector Peg/Cross Axle at the end of the 4 × 2 Angular Beams, Axle side down*

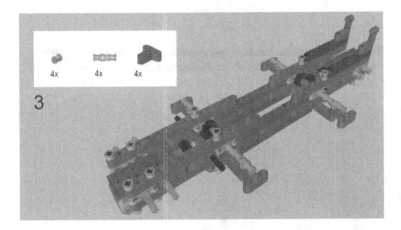

Figure 3-37. *Cap off the other end of the 6M Axles with the four Bushes. Place the Universal Joints on the 9M Axles. The four T-Beams, or 3 × 3 Angular Beams, go on the Connector Peg/Cross Axles*

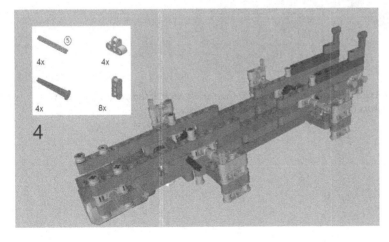

Figure 3-38. *Stack a 2 × 3 Cross Block in between two Double Cross Blocks. On one side of this stack, slide a 5M Axle and slide a 4M Axle with Stop on the other side*

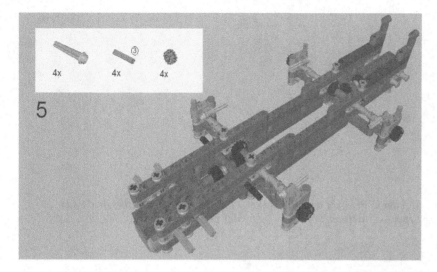

Figure 3-39. *Slide a 3M Axle through each of the round holes of the Double Cross Blocks so it fits into the ends of the Universal Joints. Place a Z12 Gear on the other end of the 3M Axles. Slide the 3M Axles with Cap through the 2 × 3 Cross Blocks, as shown, until flush*

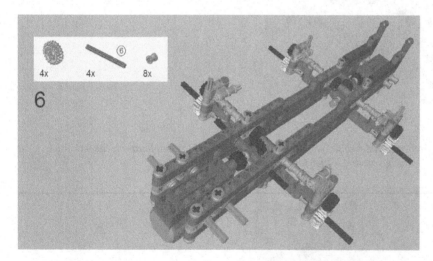

Figure 3-40. *Slide a Bush on each of the 3M Axles with a stop. Mesh a Z20 Gear with each of the other Z12 Gears, and slide a 6M Axle through that. A Bush anchors the 6M Axle and Z20 Gear into place*

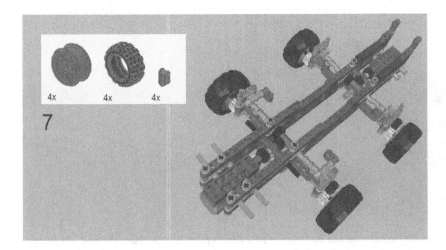

Figure 3-41. *Slide the 90 Degree Cross Blocks on the 3M Axles with Cap. Insert the Rim in each Tire and slide them into place*

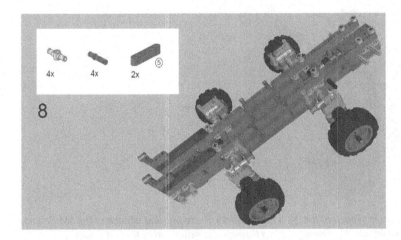

Figure 3-42. *Insert the four Double Bush with 3M into the 90 Degree Cross Blocks. Push the 3M Connector Pegs on each of the ends of the 5M Beams and insert as shown*

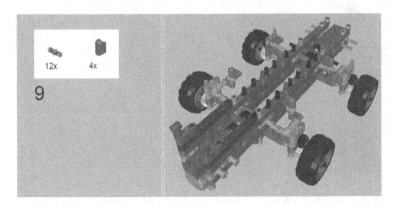

Figure 3-43. *Place a 90 Degree Cross Block on the other end of the 3M Double Bushes, then add the Connector Pegs, as shown*

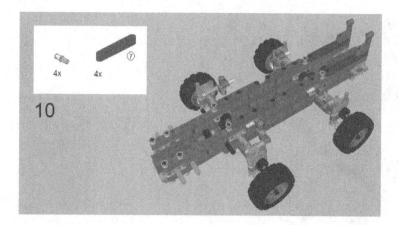

Figure 3-44. *Place the Connector Peg/Cross Axles on the 90 Degree Cross Blocks from Step 9. Snap the four 7M Beams into place*

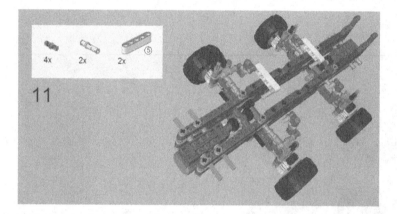

Figure 3-45. *Snap a Connector Peg in the center of each of the 2 × 3 Cross Blocks. Place one end of each of the 5M Beams on the Connector Peg/Cross Axles, and then stick the 3M Connector Peg through the other end of the 5M Beam*

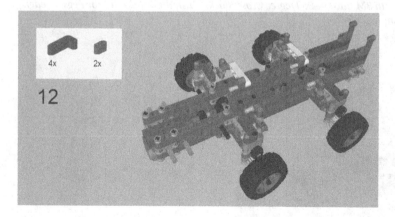

Figure 3-46. *Connect the 4 × 2 Beams to the Cross Blocks. Slide on the Cross and Hole Beams on the 3M Connector Peg in the center of the construction*

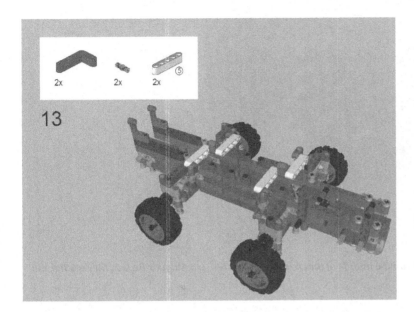

Figure 3-47. *Slide on the 5 × 3 Angular Beams, as shown, then place the Connector Pegs in the places as shown. Don't forget to snap the 5M Beams into place*

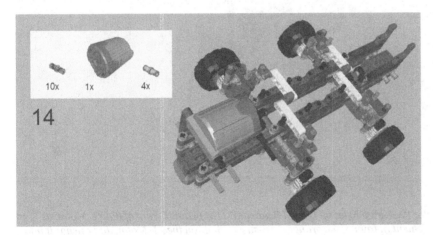

Figure 3-48. *Snap the XL Motor into place, and snap two of the Connector Pegs with Friction on the other side. The other eight Connector Pegs with Friction snap into place on the 7M Beams, as shown, with the other Four Connector Pegs going into the 4 × 2 Angular Beams*

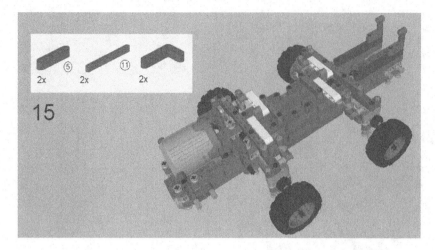

Figure 3-49. *The 11M and 5M Beams go on the middle of the creation. As for the 5 × 3 Angular Beams, they anchor the XL-Motor into place*

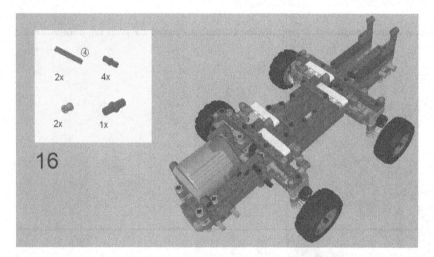

Figure 3-50. *Insert the four Connector Peg/Cross Axles on the 5M Beam and 11M Beam in front of the XL-Motor, as shown. Place the 180 Degree Angle Element in front of the orange cross-shaped hole on the XL-Motor, and anchor it into place with the Bushes and 4M Axles*

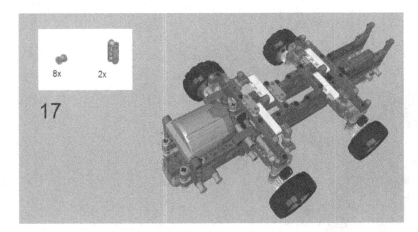

Figure 3-51. *All eight Bushes essentially anchor the 5 × 3 Angular Beams to the XL-Motor. The two Double Cross Blocks fit into the Connector Peg/Cross Axles from the previous step*

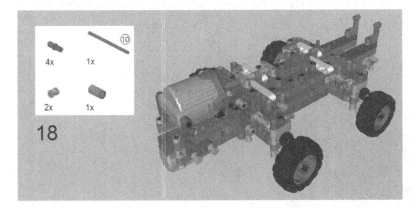

Figure 3-52. *Stick the 10M Axle into the XL-Motor. Be certain that it lines up with the cross hole of the Cross and Hole Beam and that it goes through the Tube, Bushes, Double Cross Blocks, and the 180 Degree Angle Element. Insert the Connector Peg/Cross Axles on the 5M Beam and 11M Beam on the opposite side*

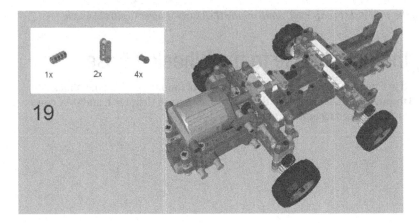

Figure 3-53. *Place the Cross-Axle Extension on the 10M Axle and the Double Cross Blocks on the Connector Peg/Cross Axles. The Connector Pegs with Bumps go in the front*

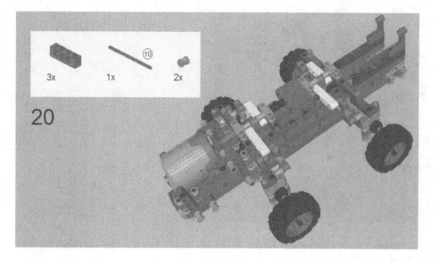

Figure 3-54. *Joint a 10M Axle to the Cross Axle Extension and center the 1 × 4 Technic Bricks. Line up the Bushes and the Cross and Hole Beam*

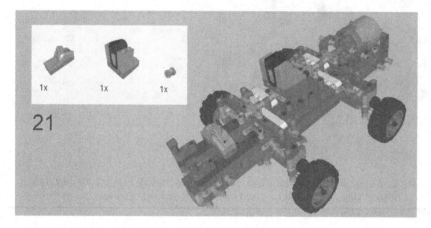

Figure 3-55. *Place a Bush at the end of the 10M Axle to cap it off. Snap on the IR-RX and Power Control Switch*

Project 3-3: Securing the Robot Base to the Wheeled Base

This next project uses the base presented in Chapter 2 (Project 2-1), and it is a way of mounting it so it has shock absorbers (11 easy steps in Figures 3-56 through 3-66). I introduced the shock absorbers in Chapter 1, and they come inh andyf ors ituationsw heny ouw anta l ittleb ounceo ny ourc reation.

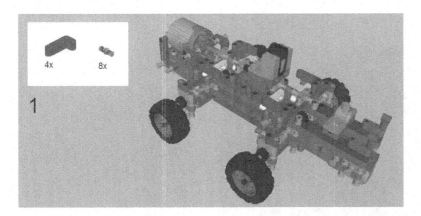

Figure 3-56. *Insert the eight Connector Pegs in the 11M Beams, just before each of its ends. Then snap the 4 × 2 Angular Beams on, as shown*

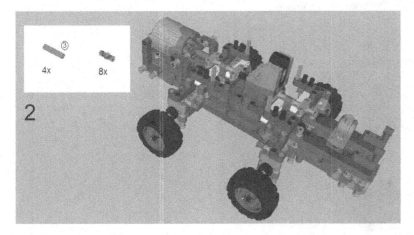

Figure 3-57. *Snap in two of the Connector Pegs on each of the 4 × 2 Angular Beams. Slide in the 3M Axle in the cross hole of the 4 x 2 Angular Beam*

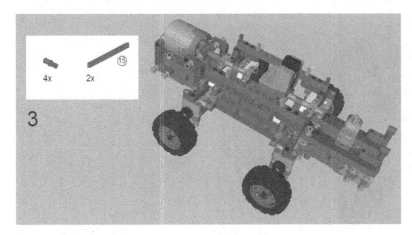

Figure 3-58. *Snap on the 15M Beams on top of the 4 × 2 Angular Beams from Steps 1 and 2. Place a Connector Peg/Cross Axle on each end of the 15M Beams*

Figure 3-59. *Insert the 3M Cross Blocks on the Axle ends, as shown*

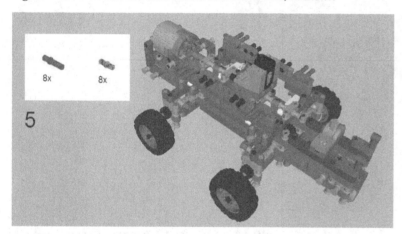

Figure 3-60. *Insert the 3M Connector Pegs on the 3M Cross Blocks on the ends and the Connector Pegs on the 3M Cross Blocks in the middle*

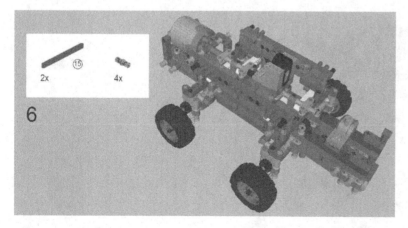

Figure 3-61. *The 15M Beams snap onto the Connector Pegs from Step 5. Insert a Connector Peg on the ends as shown*

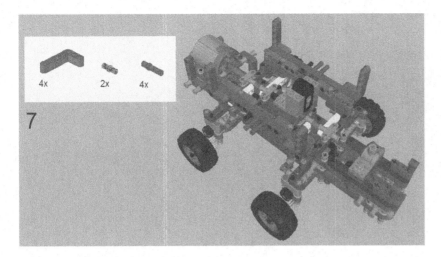

Figure 3-62. *The four 5 × 3 Angular Beams go on the ends of the 15M Beams. The 3M Connector Pegs and regular-sized Connector Pegs go in the center, as shown*

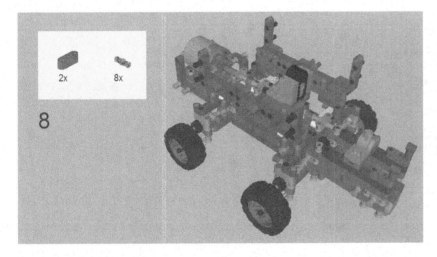

Figure 3-63. *The 3M Beam is snapped into place in the center of the 15M Beams. The Connector Pegs go on the 5 × 3 Angular Beams*

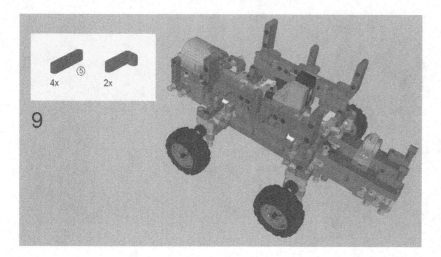

Figure 3-64. *The 5M Beams snap into place over the 5 × 3 Angular Beams. The 4 × 2 Angular Beams snap onto the ends of the 3M Connector Pegs*

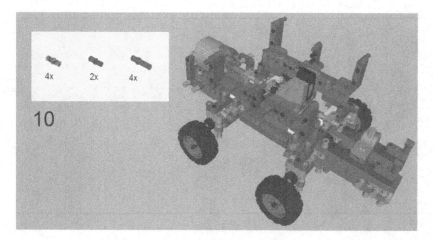

Figure 3-65. *The 3M Connector Pegs snap into place through the 5M Beams and 5 × 3 Angular Beams. The Connector Peg/Cross Axle slides into place on the 4 × 2 Angular Beam, facing opposite the 3M Connector Pegs, just like the Connector Pegs*

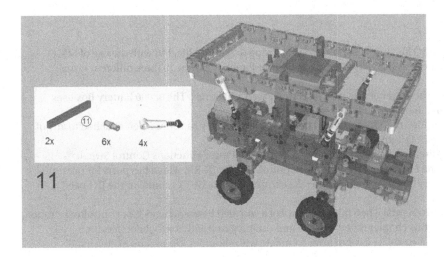

Figure 3-66. *Insert the 11M Beams on the 4 × 2 Angular Beams. Snap the six Connector Pegs into Project 2-1, as shown. The Shock Absorber pieces then join up at this point*

Powering the Robot Base

Assuming you combined Projects 3-1, 3-2, and 3-3 with Project 2-1, you need to make certain you have the motors connected to the battery.

You can see in Figure 3-67 how to hook them up. Note that the XL-Motor goes to one of the ports of the IR-RX. From here, you can take control of your creation with one of the remote controls. As for the two motors, both of them get connected to the other port of the IR-RX, stacked up. You will notice in Figure 3-67 that the switch is in place and it provides an extension so the front M-Motor can reach the IR-RX.

Figure 3-67. *Demonstration of where to hook up cables to the motors and lights to the battery box*

The IR-RX then gets connected to the battery box. The lights are also connected to it, stacked on the battery box port, with the LED bulbs plugged into the Zero Degree Elements in front.

Summary

You might not be able to create a LEGO Technic robot that you can bring to sentient life, but with the use of LEGO Power Functions, you can make it move as if it were alive. LEGO Power Functions come in three different forms: power, action, and control.

The Power Source can be the 8881 Battery Pack, which relies on six AA batteries. The 88000 Battery Box uses six AAA, and the 8878 has a lithium ion battery that charges via a DC plug-in cable.

The Action elements can be motors, the XL-Motor and the M-Motor, and they can spin when attached to any of the battery packs. LEGO also has an LED light kit that works just as easily.

Controls comes in two types: direct and remote. The direct control is a Power Functions Control Switch that is perfect for killing power for the LED light kit. For remote control, use the IR-RX, which has ports for taking control when it is attached to a battery box. It is then easy to take control with the IR-TX remote or the IR Speed Remote Control.

These Power Functions are instrumental when taking control of a wheeled base and works for controlling a robot on wheels. The creation provided in this chapter is very mobile and can turn well with four-wheel steering.

The next chapter will discuss how to create a robot arm, complete with shoulder, elbow, and even a hand.

CHAPTER 4

Designing a Robot Arm

There is an old proverb that says "Many hands make light work." In other words, if you have a big job, it helps to have as many people working on it as possible. I'm sure that whoever first conceived of the idea of a robot was probably a few people short for a big job and wanted some sort of automated hand to help finish his or her great task.

This chapter will show you how to create a robot with an arm or arms that you can manipulate by remote control. You will find that it will not be as nimble as human arms—it will not be able to toss footballs like Joe Montana or sculpt like Michelangelo—but it will be able do some pretty interesting things. Granted, most of the motions are simple, but what I have here is the basic framework to create an arm, elbow, wrist, and a hand that can clench into a fist in Project 4-1.

With the hand comes the wrist, and I will demonstrate how to create a swiveling wrist in Project 4-2 with the help of a Turntable piece. In addition to the flexible spinning wrist, Project 4-2 also shows how to create a forearm. Project 4-3 shows how to create an upper arm along with an elbow that can flex like a human elbow. Project 4-4 creates a shoulder with a ball-and-socket joint so it can swivel like the wrist and bend like the elbow. Project 4-5 will show how to join a fully completed arm with the body you created in Chapter 2. In short, you will learn to create a hand that looks human; how human and advanced you want to make it is up to you.

Project 4-1: The Hand

Have you ever noticed how adept your hand can be? One day I was eating Skittles and realized that I was sorting them by color with just one hand. This particular LEGO hand in Project 4-1 will not be that adroit, and it is really designed for gripping only. Making a hand with any bit of dexterity will require a liberal use of motors.

Of course this is not the only way to create a grip for your LEGO robot arm. You can make it with more or less fingers, but I wanted to demonstrate how to create it so it can make a good clench. All it takes is one motor and an even number of gears and you have yourself something with a strong grip. The 11 steps in Figures 4-1 through 4-11 willh elpy ouc reatet hisa rm.

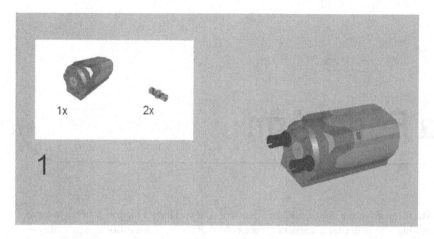

Figure 4-1. Take the M-Motor and insert two Connector Pegs in front, as shown

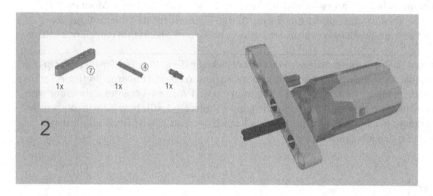

Figure 4-2. Insert a 4M Axle in the center of the M-Motor, and then place a 7M Beam on it. Note the placement of the Connector Peg/Cross Axle

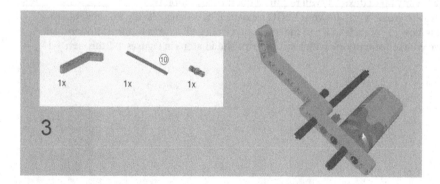

Figure 4-3. Insert a 10M Axle through the 7M Beam, then a 3 × 7 Angular Beam with a Connector Peg. Be sure to have 2M of the Axles sticking out on one side of the 7M Beam

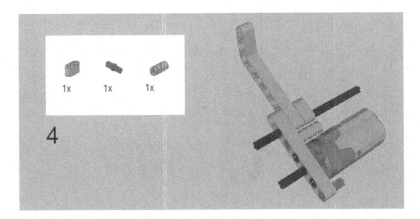

Figure 4-4. *Slide a Cross and Hole Beam onto the 10M Axle. Place a Cross-Axle Extension onto the Connector Peg Cross Axle and insert another Connector Peg Cross-Axle*

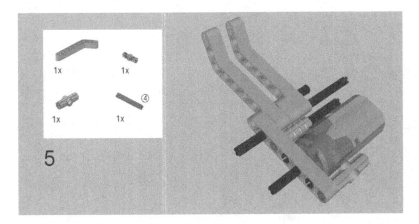

Figure 4-5. *Slide on a 3 × 7 Angular Beam with a Connector Peg. Place a 4M Axle onto the Cross-Axle Extension and a 180 Degree Angle Element onto the Connector Peg/Cross Axle*

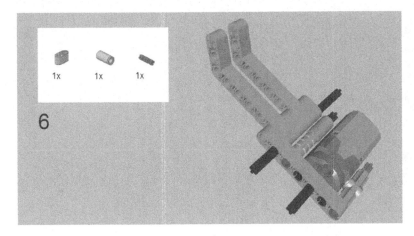

Figure 4-6. *Slide on another Cross and Hole Beam and place the Tube and 2M Axle in the positions, as shown*

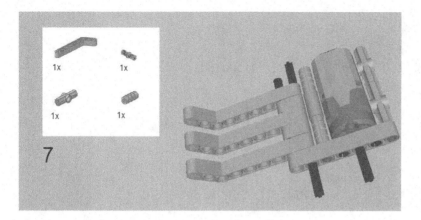

Figure 4-7. *Slide on another 3 × 7 Angular Beam with a Connector Peg. Place another Cross Axle Extension, along with a 180 Degree Element*

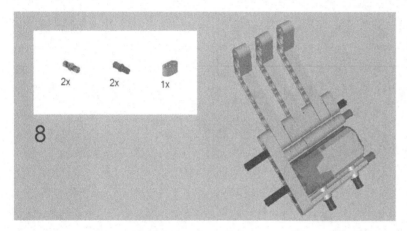

Figure 4-8. *Slide on another Cross and Hole Beam. Place two Connector Peg/Cross Axles on one side and two Connector Pegs on another*

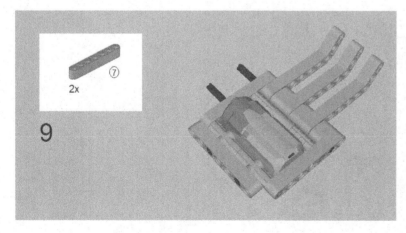

Figure 4-9. *The two 7M Beams cap off two sides of the "palm" of this hand*

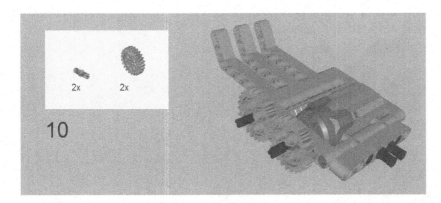

Figure 4-10. *Insert two more Connector Pegs. Slide two Gears (24 teeth) onto the Axles and be certain they mesh together*

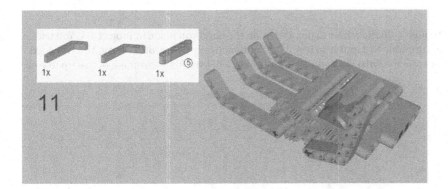

Figure 4-11. *Place a 5M Beam on the end of this palm. Insert the 3 × 7 Angular Beam next to the others. Insert a 4 × 6 Angular Beam onto the other gear to create a sort of "thumb"*

If you hook up a battery box to the M-Motor of Project 4-1, you will discover that it has the ability to grip things. It will do this quite quickly, but it might not be the right kind of grip you want. You might want to change the design of the "fingers" so it has a grip like a lobster, rather than the human hand I chose to model it after. There is more than one way to build a hand that can grip.

Project 4-2: The Wrist

This project will show you how to use a piece called a Turntable. It's worth noting its unique shape. As you can see at several angles in Figure 4-12, it is round and jagged on the edges like a gear piece, making it resemble a Reese's Peanut Butter Cup wrapper as viewed from the top. You will notice that it has two rows of three round holes on each side, and it is designed to create something that can spin 360 degrees.

Figure 4-12. *Three sides of the Technic Turntable piece, designing to spinning a creation on a controlled axis at 360 degrees*

The 16 steps in Figures 4-13 through 4-28 show how to put a wrist on the hand you made in Project 4-1. You will notice that Project 4-2 attaches this Turntable to a motor so it will spin automatically by remote control. You will need to mount an IR-RX somewhere to operate the wrist via remote control, and I will leave the placement of that up to you.

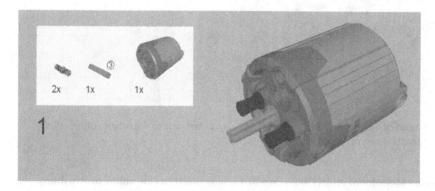

Figure 4-13. *Insert the 3M Axle into the XL-Motor and a Connector Peg on each side*

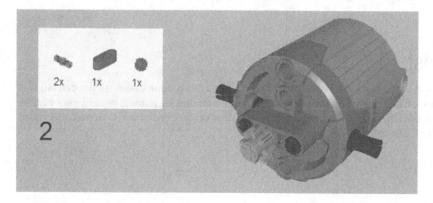

Figure 4-14. *Snap on the 3M Beam and then slide on the 8-tooth Gear. Insert a Connector Peg on the sides, as shown*

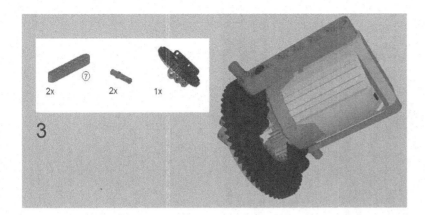

Figure 4-15. *Slide on the Turntable and make certain that the 8-tooth Gear meshes with the inner teeth of the Turntable. Snap on the 7M Beams on each side and lock it into place with the 3M Connector Pegs*

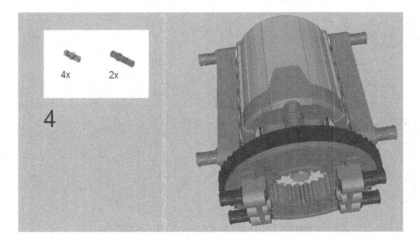

Figure 4-16. *Slide two other 3M Connector Pegs into place with the 7M Beams. Snap the Connector Pegs into place on the Turntable, as shown*

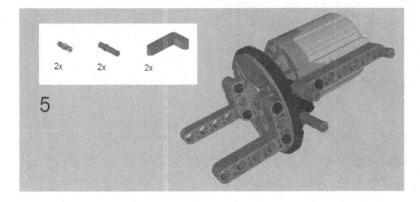

Figure 4-17. *Snap the 5 × 3 Angular Beams onto the sides of the Turntable. Note the placement of the two types of Connector Pegs on the other side of the Turntable*

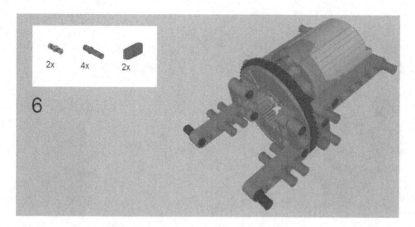

Figure 4-18. *Snap a 3M Beam onto each side. Insert the 3M and regular-sized Connector Pegs onto the 5 × 3 Beams*

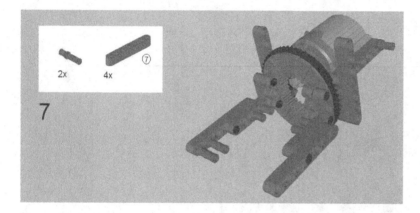

Figure 4-19. *Snap on the 7M Beams, as shown. The 3M Connector Pegs only go on one of the 7M Beams*

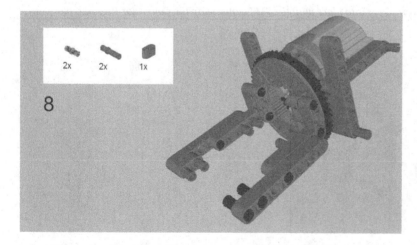

Figure 4-20. *Snap a 2M Beam onto the 3M Connector Pegs from Step 7. Insert two Connector Pegs on the opposite side of the 2M Beam. Snap in a 3M Connector Peg on each side of the 7M Beams so 1M sticks out*

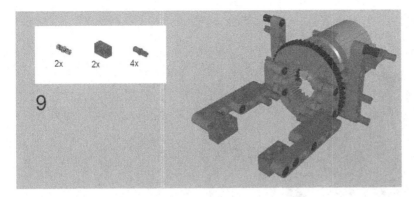

Figure 4-21. *Insert a Connector Peg on top of the 7M Beam and the Connector Peg/Cross Axles below that. Note the application of the 1 × 2 Technic Bricks, with two holes*

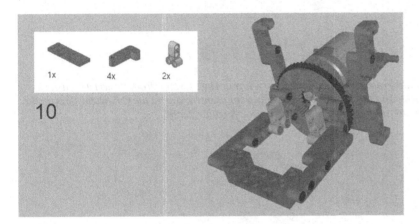

Figure 4-22. *Join the 1 × 2 Bricks with a 2 × 6 Plate. Connect the 2 × 2 Cross Blocks on the 3M Connector Pegs and insert the 4 × 2 Angular Beams in four places*

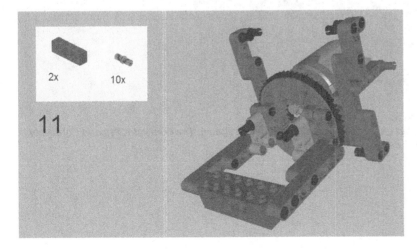

Figure 4-23. *Snap two 1 × 4 Technic Bricks onto the 2 × 6 Plate. Place Connector Pegs on the 2 × 2 Cross Blocks, as shown, and the rest of the Connector Pegs go on the 2M ends of the 4 × 2 Angular Beams*

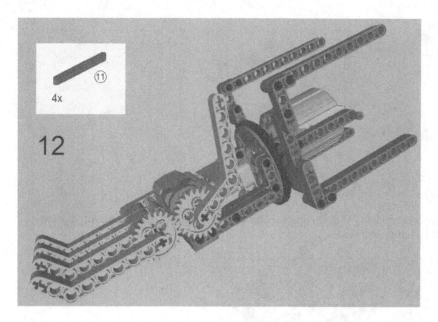

Figure 4-24. *At this point, it is time to add Project 4-1 to give the wrist a hand. The M-Motor from Project 4-1 easily snaps into place on the 2 × 6 area of bricks, and the 5M Beam on the hand snaps into place on the Connector Pegs on the 2 x 2 Cross Blocks. The four 11M Beams are snapped into place on the 4 × 2 Angular Beams*

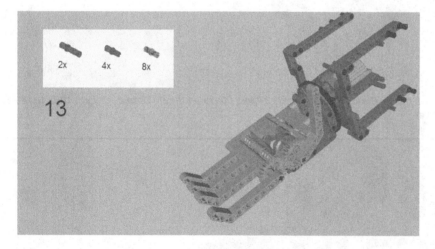

Figure 4-25. *Insert the 3M Connector Pegs on the sides on the 7M Beams, as shown. The Connector Peg and Connector Peg/Cross Axles are inserted into the 11M Beams*

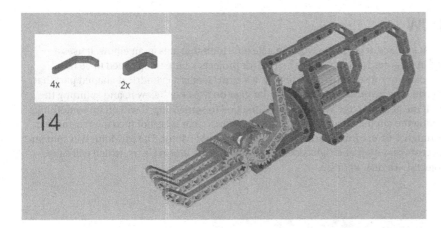

Figure 4-26. *The 4 × 2 Angular Beams join up on the sides of the 7M Beams on the XL Motor. Two Double Angular Beams go on each side, as shown*

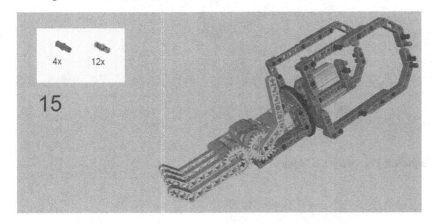

Figure 4-27. *Insert the six Connector Pegs and two Connector Peg/Cross Axles on one side, as shown, and do the same on the other side*

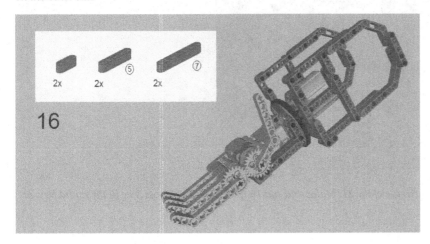

Figure 4-28. *The final step involves connecting the 3M, 5M, and 7M Beams in place to hold the forearm together*

Project 4-3: The Elbow

Now that you have the hand and wrist in place, it is time to work on a flexible joint that acts as an elbow. It uses a Worm Gear that I briefly introduced in Chapter 1, which is very helpful for projects where things need to be lifted. (I also discuss this in my previous book.) For those who need a review, a Worm Gear is a cylindrical-shaped part that is threaded like a screw, in a spiral form. The purpose is to put a circular gear on top of or below it, and spinning the Worm Gear causes the gear to turn. If the gear atop the Worm Gear is attached to something, it will lift up something, just like an elbow. By the way, spinning the circular gear atop or below the Worm Gear does not turn the Worm Gear.

The 33 steps in Figures 4-29 through 4-61 use four Worm Gears as well as an M-Motor. The M-Motor will spin an 8-tooth Gear, which will turn many gears on its left and right sides. This will turn the Worm Gears, which will cause theh and,w rist,a ndf orearmo fP rojects4 -1a nd4 -2t or ise.

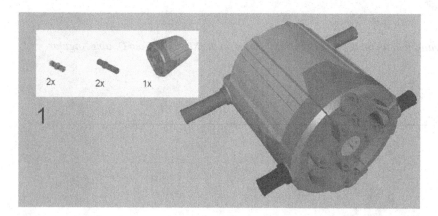

Figure 4-29. *Attach both kinds of Connector Pegs to the XL-Motor*

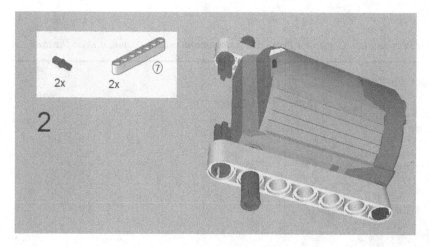

Figure 4-30. *Insert a 7M Beam on each side of the XL-Motor and snap a Connector Peg/Cross Axle in the round hole at the end*

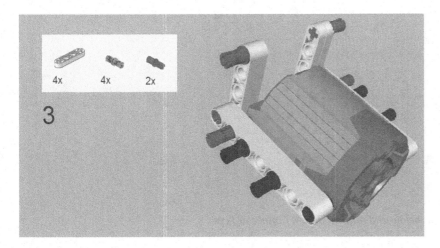

Figure 4-31. *Slide the 4M Levers onto the Connector Peg/Cross Axles from Step 2, and then slide a Connector Peg/Cross Axle onto the other end of the 4M Levers. Insert the Connector Pegs, as shown*

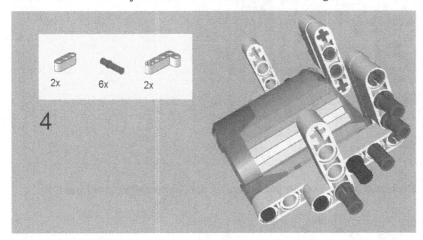

Figure 4-32. *With the help of the 3M Connector Pegs, snap a 4 × 2 Beam into place. Use the four Connector Pegs to help snap in a 3M Beam on each side*

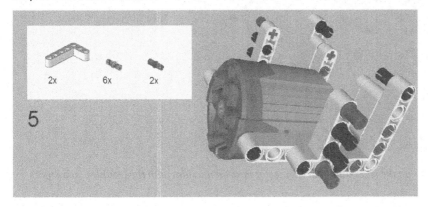

Figure 4-33. *Snap the 5 × 3 Beams into place, as shown, along with the six Connector Pegs and Connector Peg/Cross Axles*

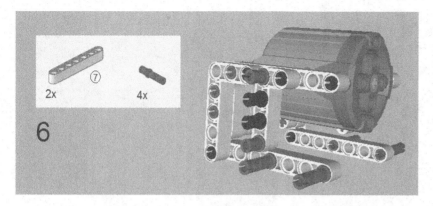

Figure 4-34. *Connect two 7M Beams below and snap on the 3M Connector Pegs*

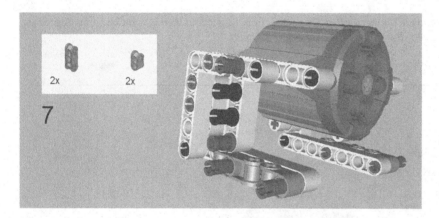

Figure 4-35. *Snap the Double Cross Blocks and 90 Degree Cross Blocks onto the 3M Connector Pegs from Step 6*

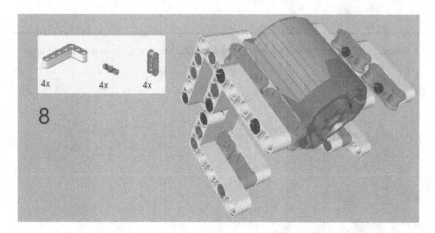

Figure 4-36. *Connect the 5 × 3 Beams on the sides and then Connector Pegs on the sides with the Double Cross Blocks*

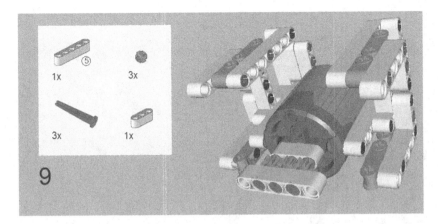

Figure 4-37. *Using the three 4M Axles with Stop, insert the 3M Beam, 5M Beam, and the three 8-tooth Gears. Make certain the gears mesh together*

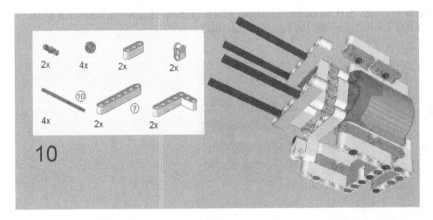

Figure 4-38. *Slide in the 10M Axles and make certain the 8-tooth Gears are on them. Place the 3M Beam, 7M Beams, and 5 × 3 Beams in the appropriate places, as well as the Connector Pegs and the 90 Degree Double Cross Blocks*

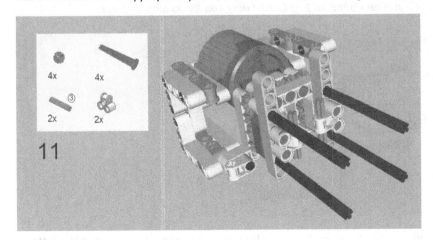

Figure 4-39. *Insert the 4M Axles with Stops with the 2 × 1 Cross Blocks and make certain the four 8-tooth Gears mesh perfectly with it. Slide the 3M Axles on the cross-hole section of the 2 × 1 Cross Blocks*

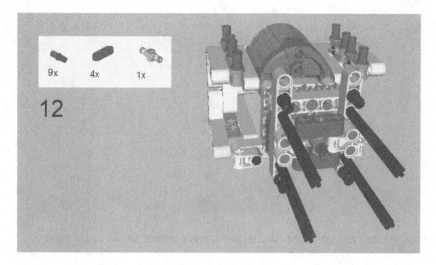

Figure 4-40. *The 3M Double Connector Peg with the Connector Peg/Cross Axle is in the middle of the sandwich of 3M Levers. Insert the rest of the Connector Peg/Cross Axles onto the Double Cross Blocks on top*

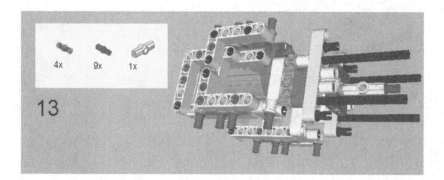

Figure 4-41. *Insert the Connector Pegs, as shown, onto the 7M Beams. The 180 Degree Element with Connector Peg/Cross Axle goes in the middle. The rest of the Connector Peg/Cross Axles go on the bottom, as shown*

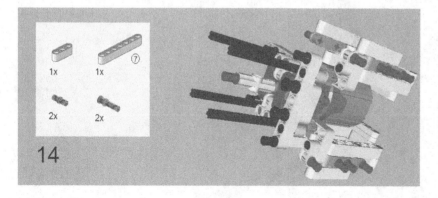

Figure 4-42. *Snap on a 7M Beam with two Connector Pegs. Insert two Friction Snaps on the 7M Beam and place a 3M Beam on the other side*

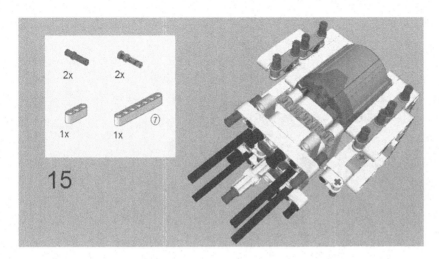

Figure 4-43. *This step looks similar to Step 14, except it uses 3M Connector Pegs*

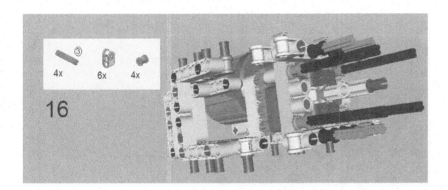

Figure 4-44. *Snap on the 90 Degree Cross Blocks, as shown. Insert the 3M Axles on the Friction Snaps. Slide the Bushes on the 10M Axles*

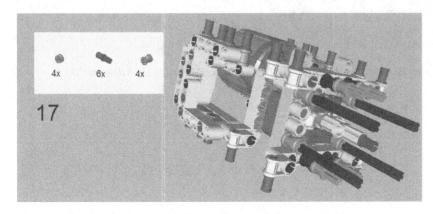

Figure 4-45. *Slide the Half Bushes onto the 10M Axles. Slide the Bushes onto the 3M Axles. Insert the Connector Peg/Cross Axles, as shown*

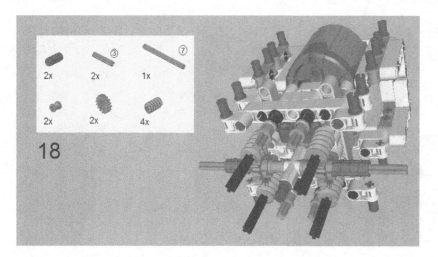

Figure 4-46. *Slide the four Worm Gears onto the 10M Axles. The rest of this step takes place in the middle. Center the 7M Axle in the middle and slide on a Bush, Z16 Gear, Cross Axle Extension, and 3M Axles onto each side*

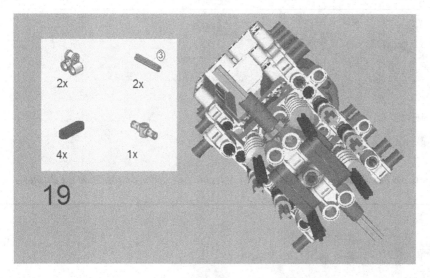

Figure 4-47. *This step is similar to Step 12 as you combine the 2 × 1 Cross Blocks, 3M Levers, 3M Axles, and 3M Connector Peg*

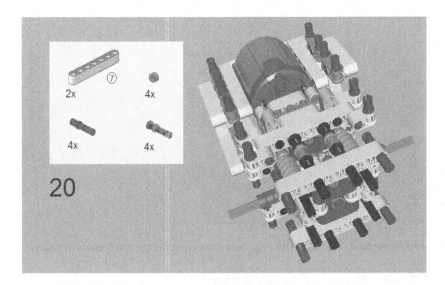

Figure 4-48. *Insert the Friction Snaps on the end of the 3M Axles. Snap on the 7M Beams on the Friction Snaps. Slide on the four Half Beams on the 10M Axles until they lock the Worm Gears into place. Don't forget the 3M Connector Pegs*

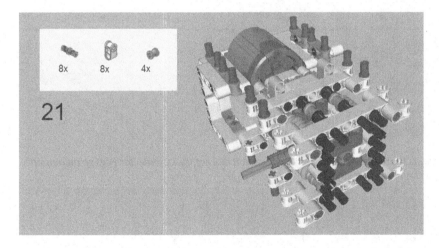

Figure 4-49. *Snap the eight Connector Pegs onto the front. Snap a 90 Degree Cross Block onto each side of the 3M Connector Pegs. Slide the Bushes onto the 10M Axles*

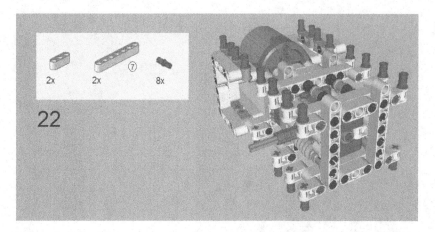

Figure 4-50. *Insert the 7M and 3M Beams on the front. The Connector Peg/Cross Axles go on the top and bottom, as shown*

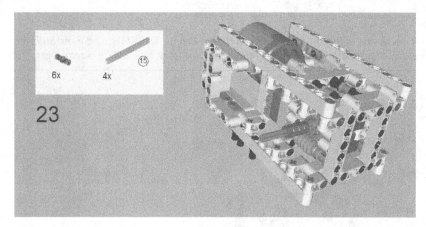

Figure 4-51. *Insert the four 15M Beams that hold the structure together. You can see that Connector Pegs go on one side, so all you have to do is mirror the effect on the other side*

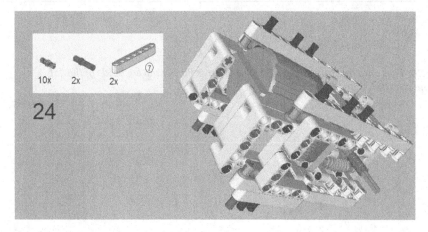

Figure 4-52. *The six Connector Pegs go on one side, while the 7M Beams go on the other, with the insertion of both types of Connector Pegs*

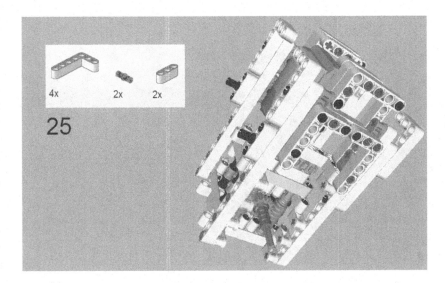

Figure 4-53. *Snap on the 3M Beams, 5 × 3 Angular Beams, and the Connector Pegs*

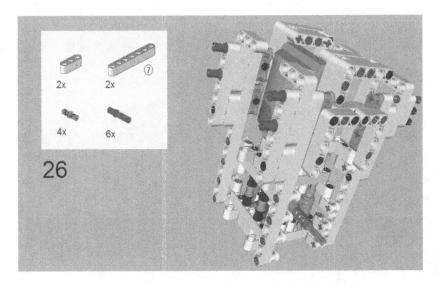

Figure 4-54. *Insert the 3M Connector Pegs and regular-sized Pegs, as shown. Snap on the 7M and 3M Beams on each side*

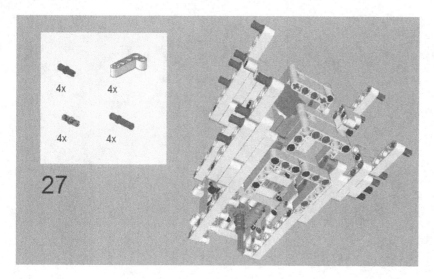

Figure 4-55. *Snap on the 3M and regular Connector Pegs. Insert the 4 × 2 Angular Beams with a Connector Peg/Cross Axles*

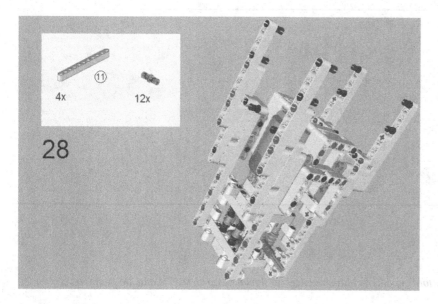

Figure 4-56. *Snap on the 11M Beams with Connector Pegs, as shown*

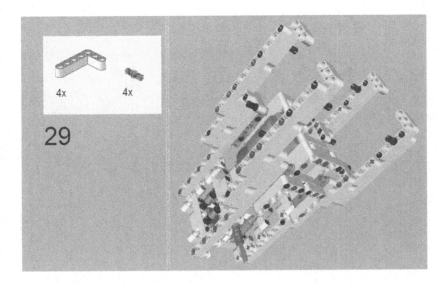

Figure 4-57. *Insert the 5 × 3 Beams with Connector Pegs, as shown*

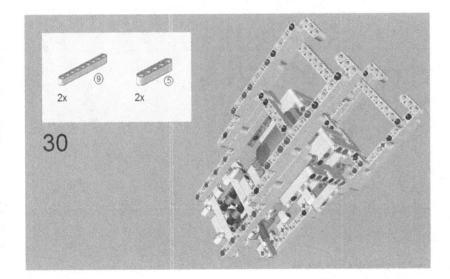

Figure 4-58. *Snap on the 5M and 9M Beams*

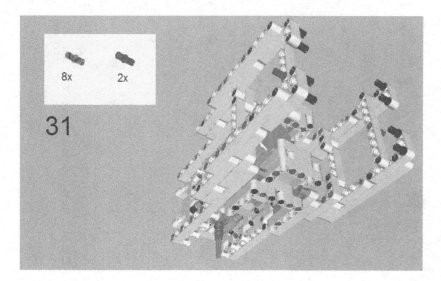

Figure 4-59. *Snap on the eight Connector Pegs plus the Connector Peg/Cross Axles*

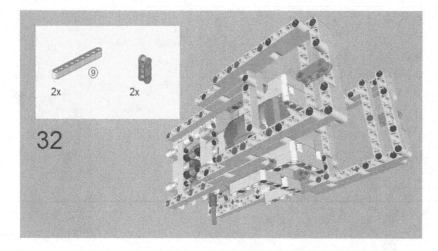

Figure 4-60. *Connect the 9M Beams and the Double Cross Blocks*

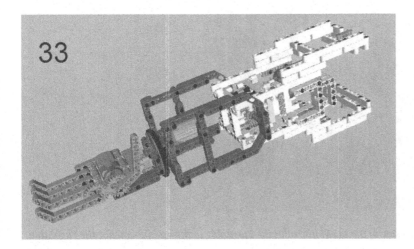

Figure 4-61. *The forearm/elbow is now complete, so go ahead and attach Project 4-2 to it. All that is required is to remove the 3M Axles from the Cross-Axle Extensions in Step 18, and then put them back in, lining the arm up as shown*

Now that you have an elbow for your forearm, you are going to want a shoulder for it as well. A shoulder joint is different from an elbow joint as it can bend and swivel as well. The next set of instructions will show you how to make that, and it is essentially combining Projects 4-2 and 4-3.

Project 4-4: The Shoulder

If iguredt hati fy oua re goingt ow antt op uta s houldero ny ourL EGOT echnicr obotb ody,t hen you will definitely need a ball-and-socket joint so it can both bend and swivel. Once again, the Turntable piece can do this, and it is possible to make it spin at 360 degrees. The 23 steps in Figures 4-62 through 4-84 will show you how to create this shoulder.

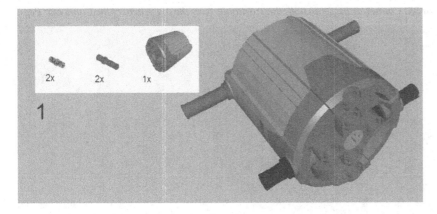

Figure 4-62. *Attach both kinds of Connector Pegs to the XL-Motor*

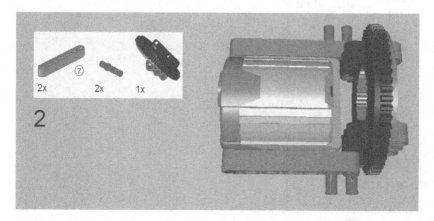

Figure 4-63. *Place the 7M Beams on each side of the XL-Motor, with the Turntable in the middle. Anchor the Turntable in place with 3M Connector Pegs*

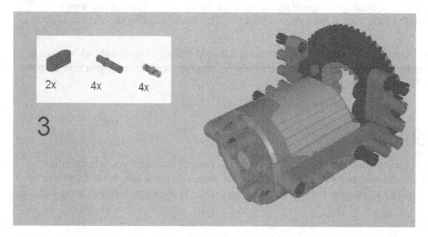

Figure 4-64. *Two 3M Beams and four 3M Connector Pegs join with the Turntable. Note the placement of the Connector Pegs*

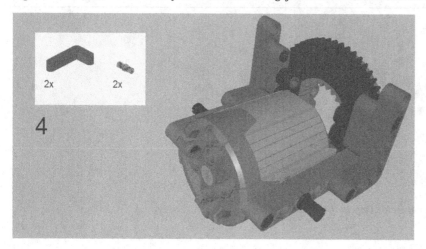

Figure 4-65. *Snap Connector Pegs into the 7M Beams, and link them with the 5 × 3 Beams*

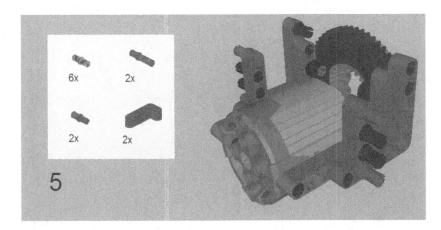

Figure 4-66. *Use a 3M Connector Peg to link the 4 × 2 Beam with the 7M Beam. Insert the six Connector Pegs and Connector Peg/Cross Axles, as shown*

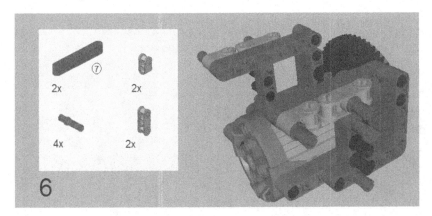

Figure 4-67. *Snap in the 7M Beam and use the 3M Connector Pegs with the 90 Degree Cross Blocks and Double Cross Blocks*

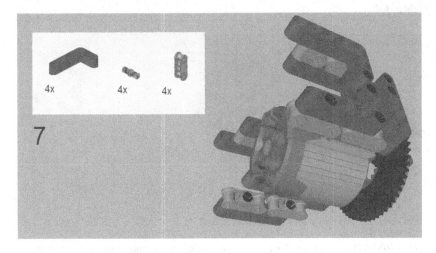

Figure 4-68. *Connect the 5 × 3 Beams to the structure, as shown, and then connect the Double Cross Blocks with the Connector Pegs*

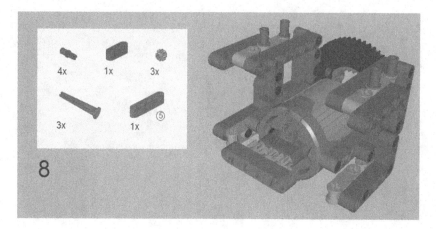

Figure 4-69. *The Connector Peg/Cross Axles go on the top of the structure. Using the three 4M Axles with Stop, insert the 3M Beam, 5M Beam, and the three 8-tooth Gears. Make certain the gears mesh together*

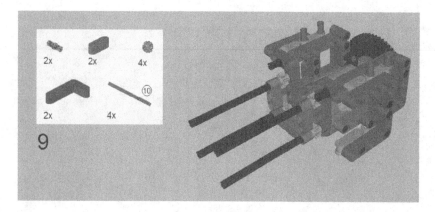

Figure 4-70. *Slide in the 10M Axles and make certain the 8-tooth Gears are on them. Place the 3M Beams, Connector Pegs, and 5 × 3 Beams in the appropriate places*

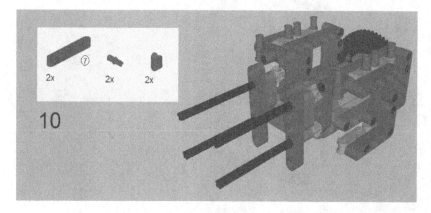

Figure 4-71. *Slide the 7M Beams onto the 10M Axles. Snap the 90 Degree Cross Blocks on top and place the Connector Peg/Cross Axles on them*

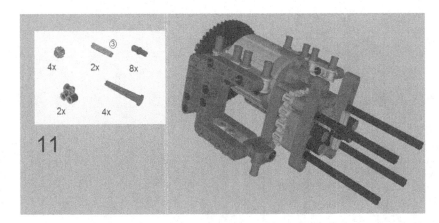

Figure 4-72. *Slide the 4M Axle with Stop and with the 1 × 2 Cross Blocks and 8-tooth Gears. Slide the 3M Axles on the cross-hole section of the 2 × 1 Cross Blocks. Insert all the Connector Peg/Cross Axles on the Double Cross Blocks*

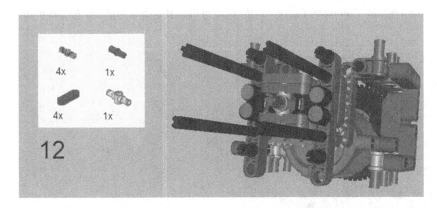

Figure 4-73. *The 3M Double Connector Peg with the Connector Peg/Cross Axle is in the middle of the sandwich of 3M Levers*

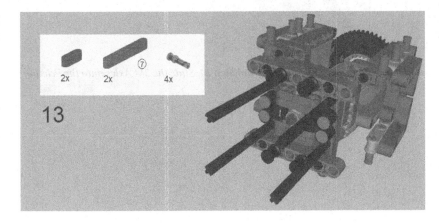

Figure 4-74. *Snap in the 7M Beams on the front. Snap in Friction Snaps with the 3M Beams on the other side*

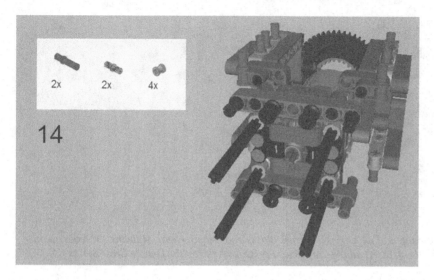

Figure 4-75. *The 3M Connector Pegs and regular Connector Pegs go on the ends of the 7M Beams. Slide the Bushes onto the 10M Axles*

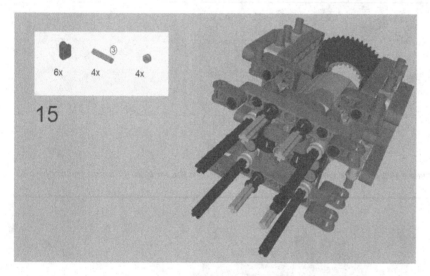

Figure 4-76. *Snap the 90 Degree Cross Blocks onto the Connector Pegs from Step 14. Slide the 3M Axles onto the Friction Snaps and slide the Half Bushes onto the 10M Axles*

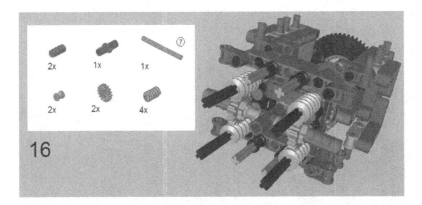

Figure 4-77. *Slide the four Worm Gears onto the 10M Axles. The rest of this step takes place in the middle. Center the 7M Axle in the middle and slide a Bush, Z16 Gear, and Cross Axle Extension onto each side*

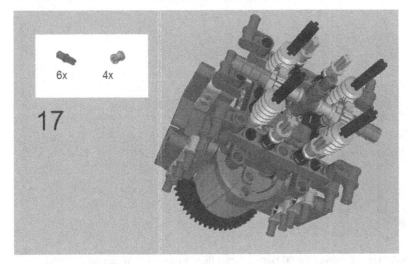

Figure 4-78. *Slide the Bushes onto the 3M Axles and insert the Connector Peg/Cross Axles on the top and the bottom through the 90 Degree Cross Blocks, as shown*

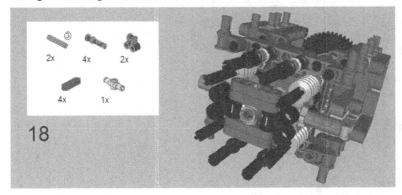

Figure 4-79. *Slide the 3M Axles onto the cross-hole section of the 2 × 1 Cross Blocks. The 3M Double Connector Peg with the Connector Peg/Cross Axle is in the middle of the sandwich of 3M Levers. Insert onto the center Connector Peg/Cross Axle. Don't forget to insert the four Friction Snaps*

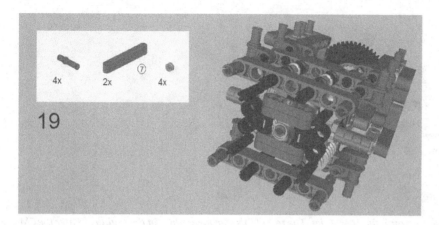

Figure 4-80. *Insert a 7M Beam on Friction Snaps and put 3M Connector Pegs on the end of them. Slide a Half Bush onto the 10M Axles*

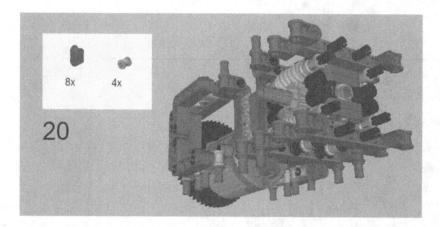

Figure 4-81. *Slide the Bushes onto the 10M Axles. Connect the 90 Degree Cross Blocks onto the 3M Connector Pegs*

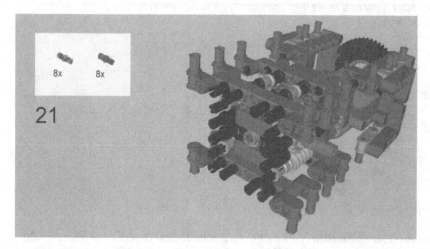

Figure 4-82. *Insert the Connector Pegs in front and the Connector Peg/Cross Axles on the top and bottom*

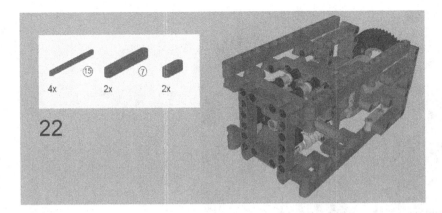

Figure 4-83. *The 3M, 7M, and 15M Beams link in many places on the structure to hold it all together*

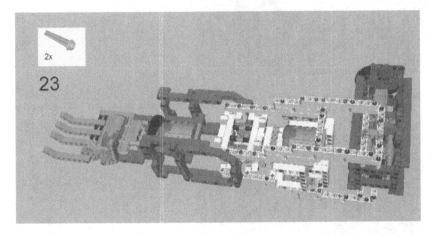

Figure 4-84. *It only takes two 3M Axles with studs to create a joined shoulder with an upper arm*

Project 4-5: Joining the Shoulder with the Body

Now that you have a shoulder, upper arm, forearm, and a hand, you will want to join it with a robot body of some type. I showed you how to construct a sample robot body in Chapter 3, so I thought I might use it as an example here. Project 4-5 will show how to join the arm to the robot body that was shown in Chapter 3 in Projects 3-1, 3-2, and 3-3.

You might also notice that I did not include anything on where to put the IR-RX or the batteries to work certain elements of the robot arm, which are completely necessary to get this to work. I will let you decide where to put those in. The 10 steps in Figures 4-85 through 4-94 will show you how to join the arm structure to the body.

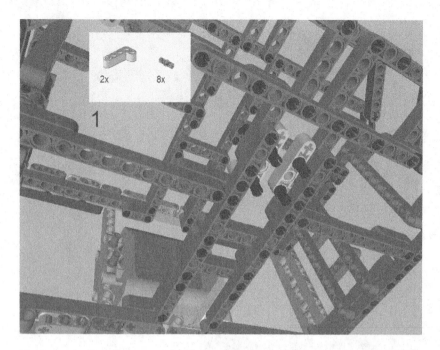

Figure 4-85. *Link the two 4 × 2 Angular Beams to the Structure with the Connector Pegs*

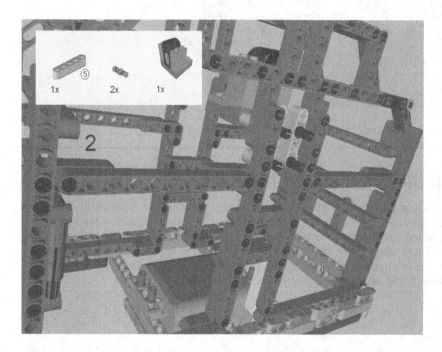

Figure 4-86. *Link the 5M Beam to the structure and then link the IR-RX with the Connector Pegs*

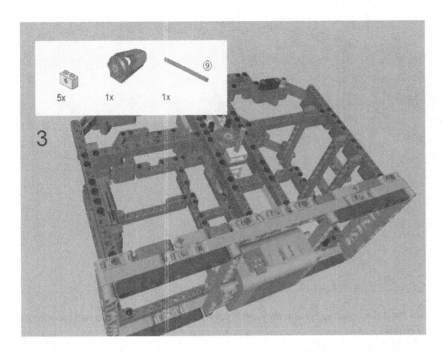

Figure 4-87. *Snap the 1 × 2 Bricks with Axle hole onto the M-Motor and slide the 9M Axle into place so it stays on the structure*

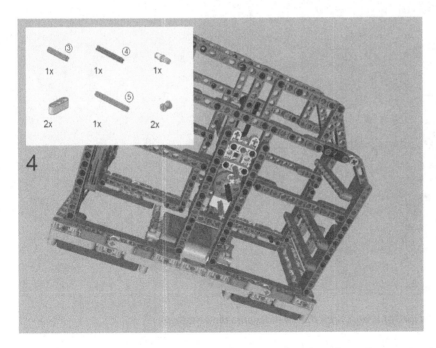

Figure 4-88. *Link the 3M Beams to the 4 × 2 Beams. Insert the Axles in their appropriate places, securing them with Bushes when needed. Add the Connector Peg/Cross Axle on the M-Motor, as shown*

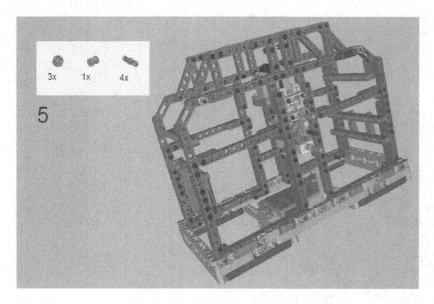

Figure 4-89. *Insert the 8-tooth Gears on the Axles, making certain they are properly meshed together. You can see where to place the Connector Pegs on the 5 × 3 Beams*

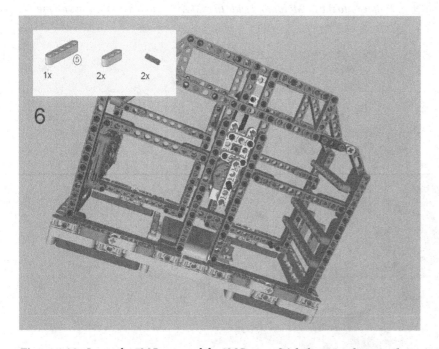

Figure 4-90. *Insert the 5M Beam and the 3M Beams. Stick the 2M Axles onto the 4 × 2 Beams*

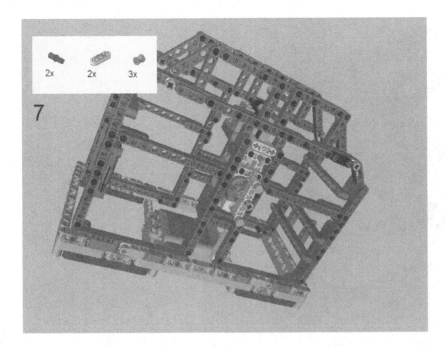

Figure 4-91. *Insert the two 3M Levers onto the 2M Axles. Cap off the Axles with the Bushes. The two Connector Peg/Cross Axles go in the center round holes of the 3M Beams*

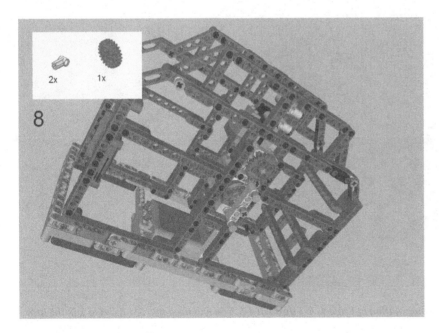

Figure 4-92. *Insert the 24-tooth Gear on the Axle and then place the Zero Degree Elements onto the Connector Peg/Cross Axles*

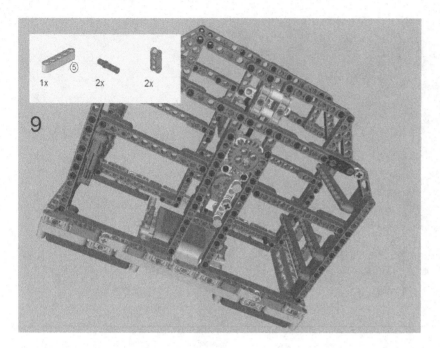

Figure 4-93. *Insert the 3M Connector Pegs with the 5M Beam. Put the Double Cross Blocks on the 3M Connector Pegs*

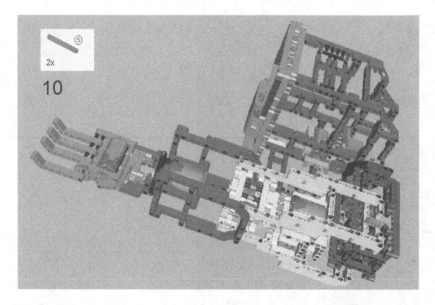

Figure 4-94. *The entire forearm, a combination of Projects 4-1 through 4-5. You will need two 5M Axles to join the arm to the body*

Once you have completed Projects 4-1 through 4-5, you may find the arm to be somewhat clunky. This is because the arm is separate from the body, and anything that is essentially an appendage in the LEGO Technic world needs some extra support. For example, my first design for the elbow joint would not stay raised very well. I had to use the gear sandwiched between two Worm Gears so it would be more stable.

In the same manner, you will discover that building a robot body with two arms will create a more balanced robot that will be more stable. If you want to put two arms on this robot body, you will discover that you might have to join some parts together. For example, if you can link the two M-Motors with a LEGO Plate, then you have a lot more stability than just one. Also, two arms will balance each other out quite well on a robot body.

Take Control with the Mega-Remote

You will also notice that it is very difficult to take command of an entire robot arm (hand, wrist, elbow, and shoulder) when the LEGO Power Functions' remote only lets you control two actions. You will need two IR-RX pieces, a battery, not to mention four motors to make the LEGO arm discussed in this chapter work. In order to take control of them, you will need two remotes.

This is where the Mega-Remote comes in. The Mega-Remote is simply two or more LEGO remote controls connected together, and though it is not an official term, I honestly can't think of a cooler or more appropriate name. You can see how easy it is to assemble in Figure 4-95.

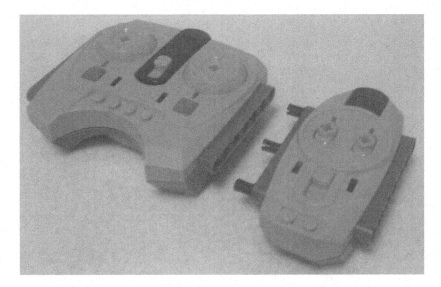

Figure 4-95. *This is a way to attach two remotes together for a Mega-Remote*

Now remember when I first introduced these remotes in Chapter 3, I mentioned that these have four different channels? You will need to make certain that your IR-RX remote controls are on different channels so you can take control of separate functions. Connect all your controllers to the red or blue areas, and then plug the IR-RX remotes into the battery.

Summary

Making a LEGO arm is relatively easy if you break down an arm into its component parts. For example, you can start with the shoulder, and even make it spin at 360 degrees with the Turntable part. The wrist can also use a Turntable for more flexibility.

As for the elbow or other hinge joints, the use of the Worm Gear is one of the best ways to get a pivot on a hinge. The hand can be very intricate if you want it to be, but I highly recommend using two gears and one motor to create a grip.

Depending on what you want your robotic arm to do, you might want to connect two or more remotes together with what I call a Mega-Remote. This will allow you to take control of several functions at once and give your robot arm more control.

The next chapter will show how to create an arm that can extend, not to mention a base that will rise like a scissorl ift.

CHAPTER 5

■ ■ ■

Creating Robots with Extensions

Some of us love the idea of robots that can stretch. I am not talking about superheroes like Plastic Man or Mr. Fantastic, who have bodies made of rubber. I am talking about robots who have extendable limbs and other parts.

For example, there is Machine Man. If you never heard of Machine Man, I can't blame you. Machine Man was a character created for a comic book series based on the famous Stanley Kubrick film *2001: A Space Odyssey*. He was created by Jack Kirby, a comic book writer/artist, who also helped to create a number of very famous Marvel comics characters like The Fantastic Four, X-Men, and the Hulk. Apparently, Machine Man was pretty popular, and he was made into a Marvel character along with all the other famous Marvel characters created by Jack Kirby and Stan Lee. Machine Man had the power to extend his arms and legs like telescopes, an ability similar to the cartoon character Inspector Gadget.

The two projects in this chapter will show you how to create robots with extensions on them. The first project uses a rack-and-pinion method, which involves a gear and a rack to extend out from where it normally wouldn't, and it can be used on arms, legs, waists, and other places you need extension. The second project also involves a rack-and-pinion method, but it uses a method that can be seen on most scissorlifts.

Your LEGO Technic robots might not have telescoping limbs like Machine Man or Inspector Gadget, but there are ways to create extendable parts on a robot that I will detail here. Why? Because it's downright nifty, that's why. If you don't believe me, look at Figure 5-1.

Figure 5-1. *A nonrobotic example of extendable arms*

■ **Note**   Before beginning this chapter's projects, refer to Appendix A for a complete list of required parts.

Project 5-1: Rack-and-Pinion Extension

If you are not familiar with the rack-and-pinion method, it is a simple mechanism in which a rack (a toothed section) intersects with a pinion (or gear). It is set up so that when the pinion is turned, the rack moves. In my first book, *Practical LEGO Technic*, I discussed how the rack-and-pinion method can be used to steer a vehicle. In this project (Figures 5-2 through 5-34), I'll show you how to create a system where a rack and a pinion can be used to raise something, in a manner that is very similar to a forklift an elevator without a cable.

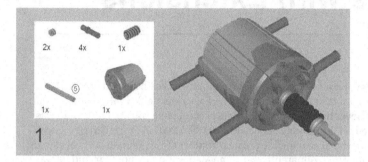

Figure 5-2. *Stick a 5M Axle into an XL-Motor. Place a Worm Gear in between two Half Bushes on the Axle, leaving 1M on the end. Insert the four 3M Connector Pegs into the round holes on the XL-Motor, as shown*

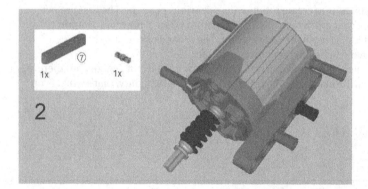

Figure 5-3. *Slide a 7M Beam onto one of the 3M Connector Pegs, as shown, and then place a Connector Peg on this*

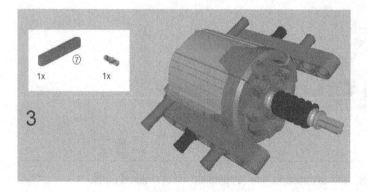

Figure 5-4. *This step is just like Step 2, but the 7M Axle and Connector Peg go on the opposite side*

Att hisp oint,w eb egina n ewp rojectt hatw ille ventuallyj oinu pw itht hem ainp roject.

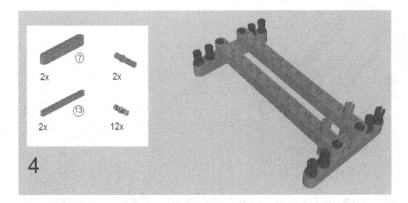

Figure 5-5. *Attach three Connector Pegs on the ends of the 7M Beams and then place two 13M Beams on top of that. Place the 3M Connector Pegs in the 13M Beams, as shown*

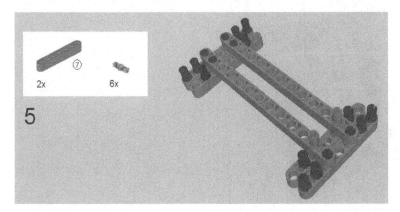

Figure 5-6. *Two more 7M Beams join the construction. One side uses four Connector Pegs while the other side uses two Connector Pegs and the 3M Connector Pegs from Step 4*

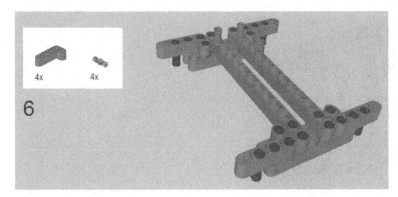

Figure 5-7. *Four 4 × 2 Beams go on as shown, with a Connector Peg inserted on the bottom as shown*

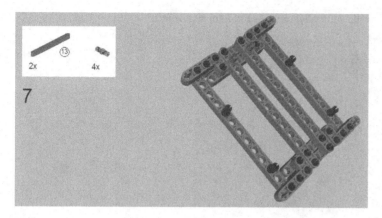

Figure 5-8. *Snap the 13M Beams onto the four Connector Pegs from Step 6. Four more Connector Pegs go in place on the 13M Beams, as shown*

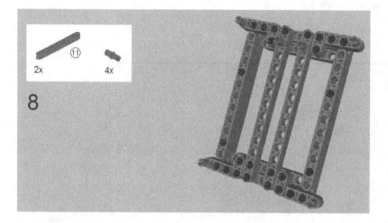

Figure 5-9. *Snap two 11M Beams onto the Connector Pegs from Step 7. Place a Connector Peg/Cross Axle at the end of the 4 × 2 Beams on the cross-shaped end*

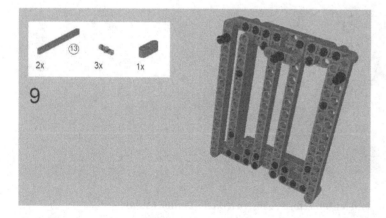

Figure 5-10. *Connect two 13M Beams to the other side, onto the Connector Peg/Cross Axles from Step 8. Stick the 3M Beam onto the 3M Connector Pegs and then stick on the three Connector Pegs, as shown*

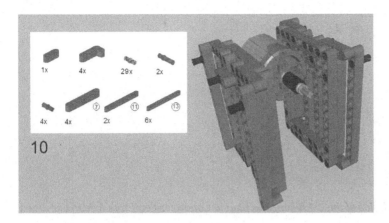

Figure 5-11. *This step may look complicated due to the number of parts, but it actually is quite simple. Just repeat Steps 4-9, which will create another construction like in Step 9. All that's left to do is "sandwich" the creation from Steps 1-3 in between the two identical creations. Note the placement of the XL-Motor*

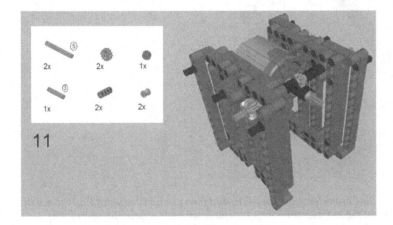

Figure 5-12. *This is a more complicated step. You start by centering an 8-tooth Gear in a 3M Axle, and placing that in the center of the Worm Gear, then put a Cross Axle Extension on each side of that. On each side, place a 5M Axle with a Bush and Z12 Gear, with 1M Axle sticking out, as shown*

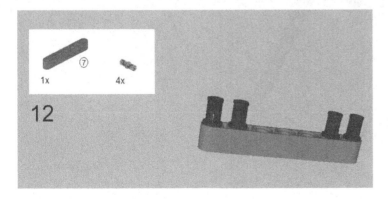

Figure 5-13. *Once again, this is another step that is separate from the rest of the creation. It starts with a 7M Beam and then applies four Connector Pegs*

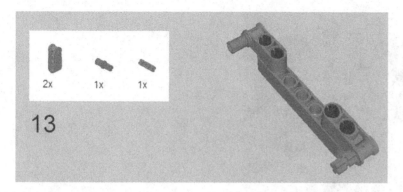

Figure 5-14. *Place a 3M Cross Block on each side of the 7M Beam, and place a Connector Peg/Cross Axle in one and a 2M Axle in the other*

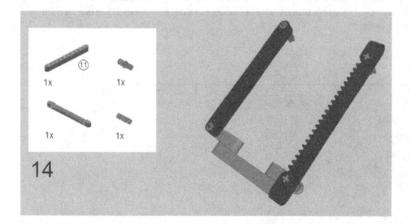

Figure 5-15. *Place an 11M Beam on the end of the Connector Peg/Cross Axle, and then place a Connector Peg/Cross Axle on the end of that. Place a 13M Rack on the end of the 2M Axle from Step 13, and place a 2M Axle on the end of that*

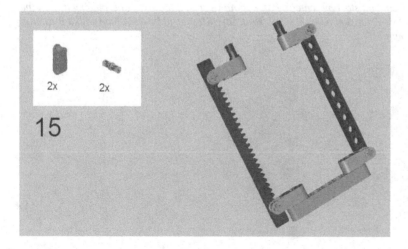

Figure 5-16. *Attach a 3M Cross Block on the ends of the 11M Beam and 13M Rack and place a Connector Peg in each*

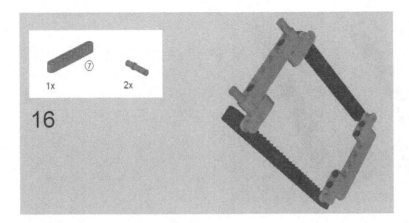

Figure 5-17. *Place a 3M Connector Peg in each of the 3M Cross Blocks, and then connect it all together with a 7M Beam*

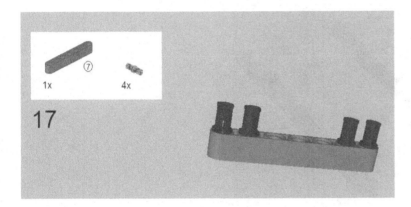

Figure 5-18. *Once again, this is another construction separate from the rest, and it is identical to Step 12*

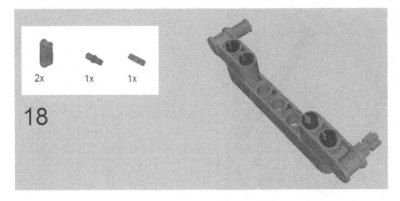

Figure 5-19. *This step looks similar to Step 13, but the Connector Peg/Cross Axle and 2M Axle stick out from the 3M Cross Blocks on the opposite side. This is quite important*

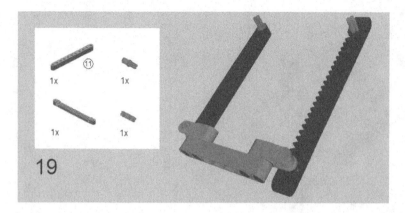

Figure 5-20. *An 11M Beam and a 13M Rack go on at this point. Note the direction that the teeth on the rack are facing. Also note the position of the Connector Peg/Cross Axle and the 2M Axle*

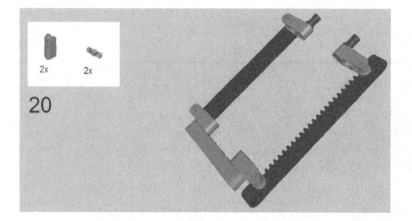

Figure 5-21. *The 3M Cross Blocks go on the Connector Peg/Cross Axle and the 2M Axle. Two Connector Pegs go on them*

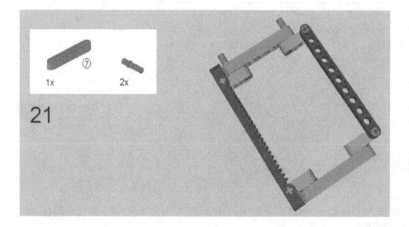

Figure 5-22. *Place a 3M Connector Peg in each of the 3M Cross Blocks and then connect it all together with a 7M Beam*

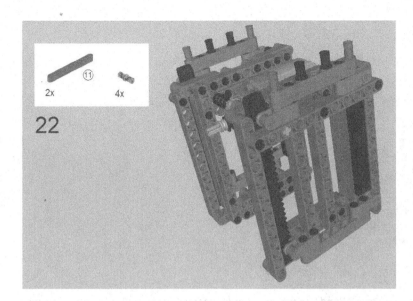

Figure 5-23. *The two sections then go on the side of the constructions so they mesh with the Z12 Gears. The 11M Beams go on the side to hold the racks in place. Remember to snap four Connector Pegs on top*

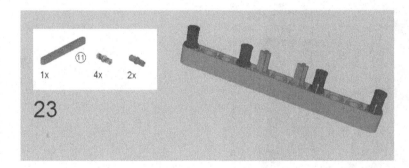

Figure 5-24. *This is another separate construction. Start with an 11M Beam and place the Connector Pegs and Connector Peg/Cross Axles on, as shown*

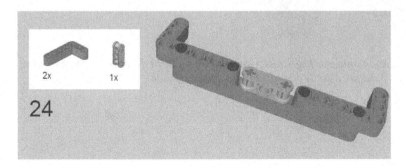

Figure 5-25. *Place the two 5 × 3 Beams and Double Cross Block on top, as shown*

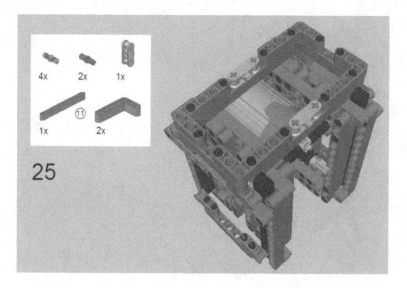

Figure 5-26. *Take the separate creation from Steps 23 and 24 and snap it onto the top as shown. Build another one just like it and snap it onto the top on the other side*

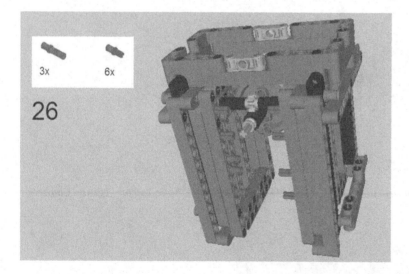

Figure 5-27. *Insert the 3M Connector Pegs and Connector Peg/Cross Axles on the left side in the round holes, as shown. In all but one case, the Connector Peg/Cross Axles are directly across the pieces on the left side*

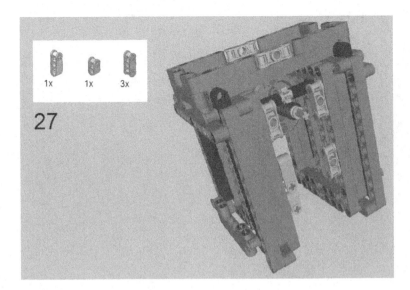

Figure 5-28. *The Double Cross Blocks slide onto the Connector Peg/Cross Axles from the previous step, while the 3M Cross Block and 90 Degree Cross Block snap onto the 3M Connector Pegs*

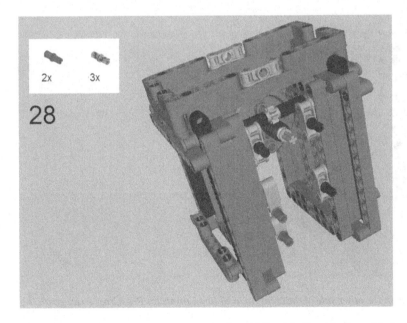

Figure 5-29. *Insert the Connector Pegs in the middle of the Double Cross Blocks. Insert the Conenctor Peg/Cross Axles into the 3M Cross Block and 90 Degree Cross Block*

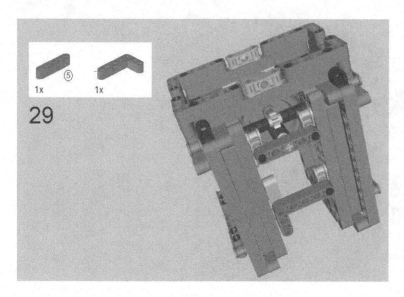

Figure 5-30. *Snap the 5M Beam and 5 × 3 Beam in place, as shown*

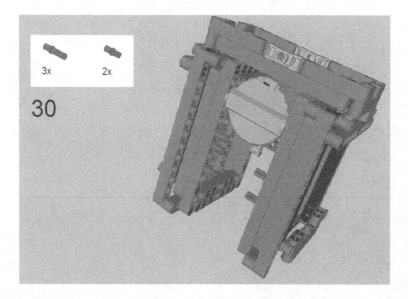

Figure 5-31. *Three 3M Connector Pegs go on the other side of the creation, and two Connector Peg/Cross Axles go on the opposite side*

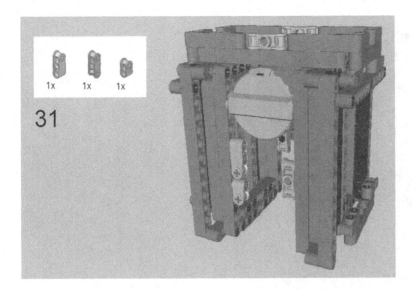

Figure 5-32. *Note the placement of the 3M Cross Block, 90 Degree Cross Block, and Double Cross Block*

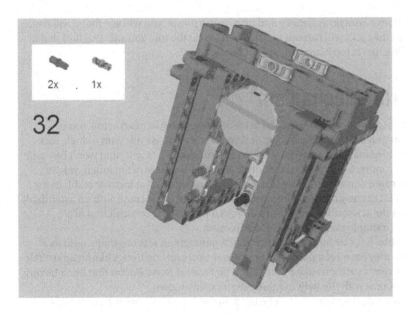

Figure 5-33. *Insert the Connect Peg and Connector Peg/Cross Axles, as shown*

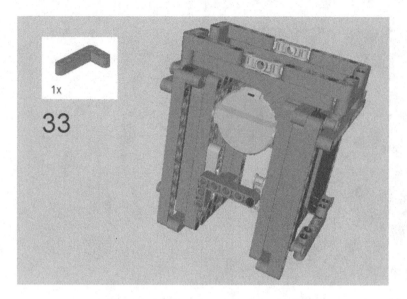

Figure 5-34. *Snap in the 5 × 3 Beam to complete the project*

If you connect the motor to a LEGO Power Functions Battery and turn on the power, you will see that it will have lift. You don't want to leave it turned on, as the racks will run out of room and jam at the top. You will also find that it can bear a load, but not too heavy. Anything that is too heavy will push it down.

Project 5-2: The Scissorlift Mechanism

You will find that the rack-and-pinion type of extension is very helpful in cases where you just need a little more telescopic extension. You could put it on an arm or neck, for example, and you could even use this with a single rack and pinion, provided you have some method of making sure what you are extending is on a "track" that won't be easily derailed. In other words, just make certain your rack and pinion is solidly built and doesn't carry too much weight.

You may discover that you want your extension to carry a lot of weight, like an entire robot torso or head. In my previous book, I discussed how to create a lot of construction equipment. One of them was a crane with an extendable end, but one vehicle that I failed to cover is the scissorlift. I'm going to discuss it now, as the construction of a scissorlift can take much more weight than a simple rack-and-pinion mechanism.

If you have never seen a scissorlift before, it is essentially a pantagraph. A pantagraph was originally used as a device to copy and scale diagrams, but you may have seen pantagraphs as extension arms on things like an adjustable wall-mounted mirror. I see pantagraphs in many cartoons like a mallet in *Who Framed Roger Rabbit* that has a boxing glove in it so it can punch things from a distance with the help of a pantagraph mechanism.

The scissorlift you will create in Figures 5-35 through 5-63 can be used on the "waist" of your robot, and it is perfect when you want your mechanical creation to have some more height. Like Project 5-1, it uses a rack-and-pinion system, but in a different manner.

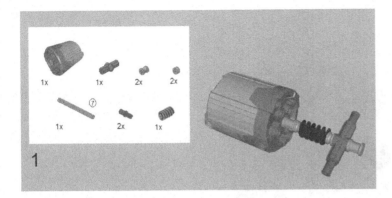

Figure 5-35. *Start with an XL-Motor and put on the 7M Axle. Put on the Half Bushes, Worm Gear, Bushes, and the 180 Degree Angle element. Place the Connector Peg/Cross Axles on the ends of the 180 Degree Angle Element*

Unfortunately, I can only show one side of the construction due to the two-dimensional nature of illustrations. For the next few steps, you will need to be symmetrical. If you put parts on one side, you will also need to put them on theo thers ide.

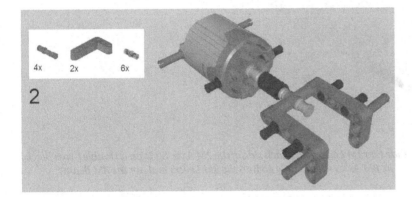

Figure 5-36. *Place two 3M Connector Pegs on the sides of the XL-Motor. Place the 5 × 3 Angular Beams on the ends of the Connector Peg/Cross Axle, and put two of the 3M Connector Pegs on it. The other Connector Pegs go in the 5 × 3 Angular Beams and XL-Motor, as shown*

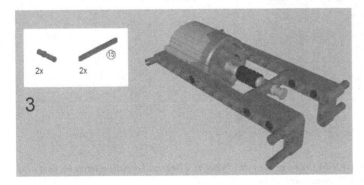

Figure 5-37. *Put the Connector Pegs on the 5 × 3 Beams and then put on the 15M Beams to firmly hold the structure together*

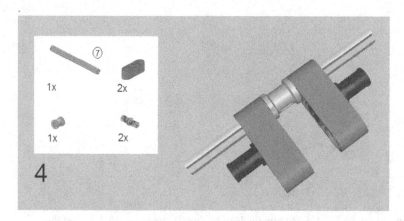

Figure 5-38. *In this particular step, you will need to begin a separate construction that you see here. Slide a Bush into the middle of a 7M Axle and slide 3M Beams onto each side. Insert a Connector Peg in the center of each of the 3M Beams*

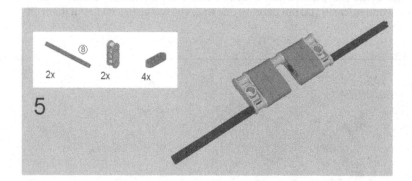

Figure 5-39. *On the section from Step 4, slide two 3M Levers onto each side of the 7M Axle. Slide on a Double Cross Block, and then slide the 8M Axle onto each side. The 8M Axles should only go into the 3M Levers and not the 3M Beams*

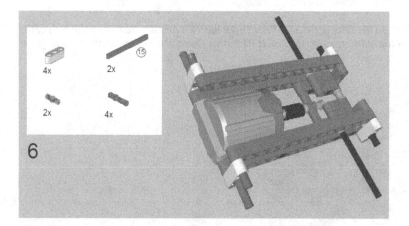

Figure 5-40. *The 3M Beams go on each corner, and a 15M Beam goes on the bottom. The small creation from Steps 4 and 5 is sandwiched in between the 15M Beams, and it will be secured later*

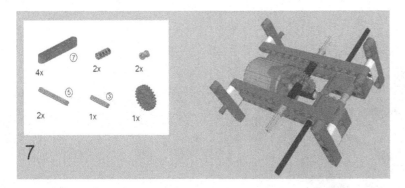

Figure 5-41. *Take the 5M Axles, Cross Axle Extensions, Bushes, 3M Axle, and 24-tooth Gear and place them so the 24-tooth Gear meshes with the Worm Gear. A 7M Beam goes on each corner as shown*

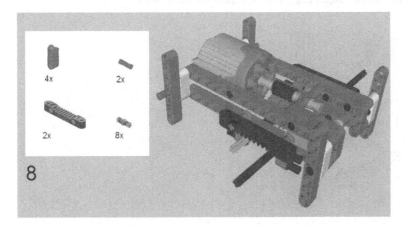

Figure 5-42. *Slide the 3M Cross Block onto the 8M Axle, and then slide on the 7M Rack. Put a 2M Axle on the other side of the 7M Rack, and then put the 3M Cross Block on the other side. Place the Connector Pegs onto the 3M Cross Blocks*

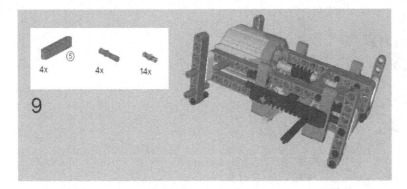

Figure 5-43. *Place the 5M Beams on the 3M Cross Blocks and put Connector Pegs on the 5M Beams. Place the 3M Connector Pegs, as shown, and put the remaining Connector Pegs onto the 7M Beams*

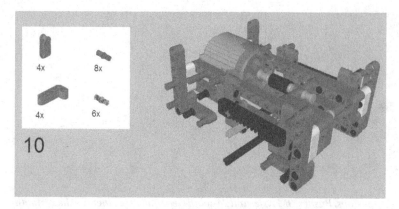

Figure 5-44. *Place the 3M Cross Blocks on the 5M Beams, and put the Connector Peg/Cross Axles there. Put the 4 × 2 Angular Beams on the corners with Connector pegs and Connector Peg/Cross Axles*

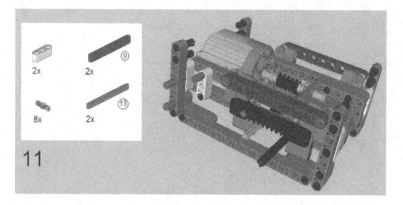

Figure 5-45. *Place the 9M Beam over the 7M Rack and put the 13M Beam below it. The 3M Beams go on the sides, and two Connector Pegs go on the top corners*

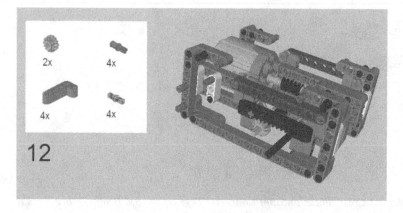

Figure 5-46. *Slide the Z12 Gears onto the 5M Axles to mesh with the 7M Rack. The 4 × 2 Angular Beams go in the corner with the Connector Pegs and Connector Peg/Cross Axles*

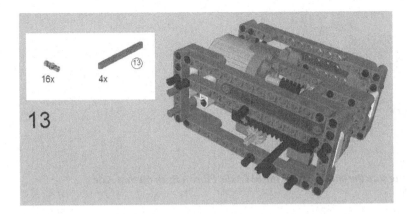

Figure 5-47. *Place the 13M Beams on the top, and then place the Connector Pegs on, as shown*

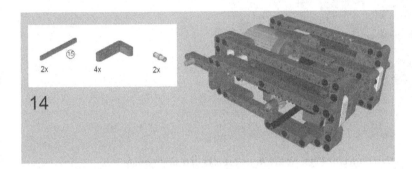

Figure 5-48. *Place the 15M Beam on the middle of the construction, and that will lock the racks into place. The 5 × 3 Beams go in as follows, with a Connector Peg without Friction*

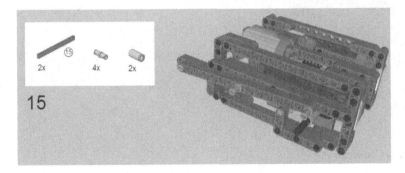

Figure 5-49. *At this point, you will be creating the scissorlift portion. Place a Tube on each side of the 8M Axles and two Connector Pegs on the 15M Beams*

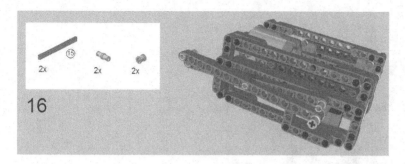

Figure 5-50. *Put on another Beam, making sure the end of it is on the 8M Axle. Place a Bush on that Axle, and repeat these steps for the opposite side*

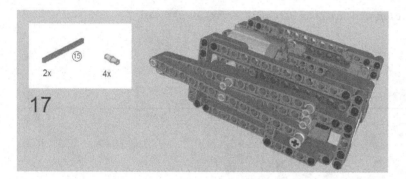

Figure 5-51. *Add another 15M Beam with the connector Pegs. Notice how it stacks at a slight angle*

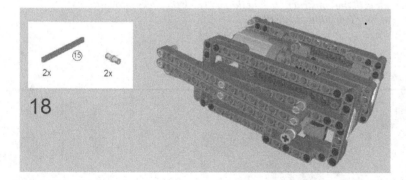

Figure 5-52. *Add another 15M Beam and a Connector Peg*

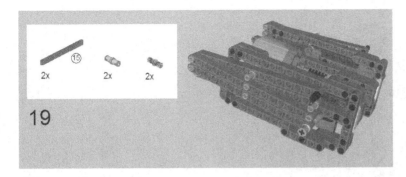

Figure 5-53. *Add another 15M Beam with some Connector Pegs*

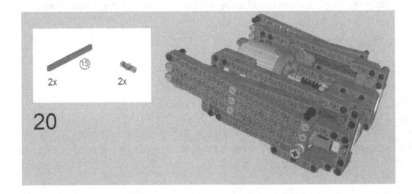

Figure 5-54. *With another 15M Beam and Connector Peg, the scissorlift part of the construction is complete*

At this point, there is no "mirroring" on the other side of the construction.

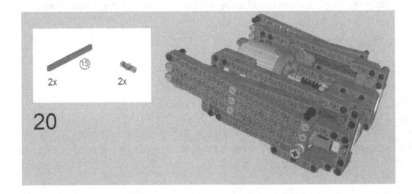

Figure 5-55. *Place a Cross and Hole Beam on one side, and then put on a 4 × 2 Angular Beam with Connector Peg and Connector Peg/Cross Axle*

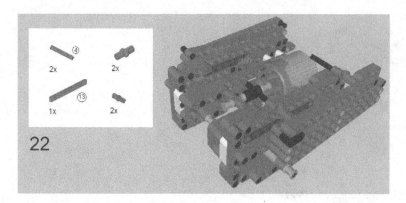

Figure 5-56. *Attach a 13M Beam along with two 180 Angle Elements, 4M Axles, and two Connector Peg/Cross Axles*

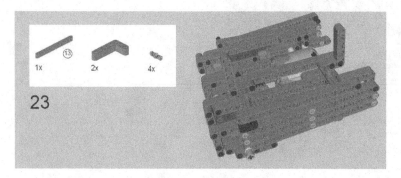

Figure 5-57. *Place two 5 × 3 Beams on the Axles with the 13M using the Connector Pegs*

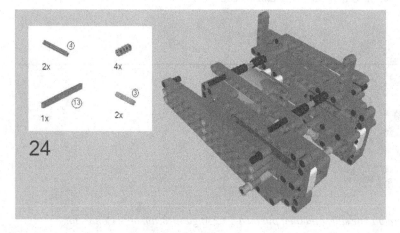

Figure 5-58. *Put the Cross Axle Extensions on the 4M Axles, then put 3M Axles on them. Put the Cross Axle Extensions on the 3M Axles, and put the 4M Axles on those. Put the 13M Beam here*

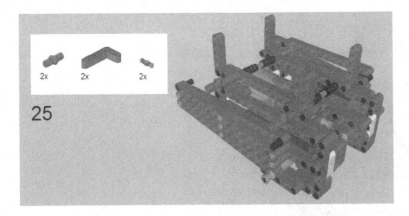

Figure 5-59. *Place the 5 × 3 Beam, Connector Pegs, and 180 Angle Elements here*

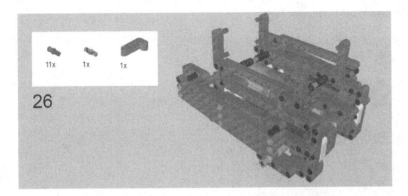

Figure 5-60. *Most of the Connector Peg/Cross Axles go on the 5 × 3 Angular Beams. The rest is on the 180 Degree Angle Elements and the 4 × 2 Angular Beam made to look like the other side with the addition of a Connector Peg*

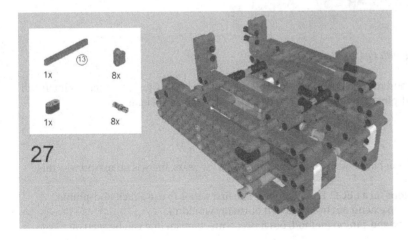

Figure 5-61. *Add a 13M Beam and a Cross and Hole Beam. Add eight 90 Degree Cross Blocks*

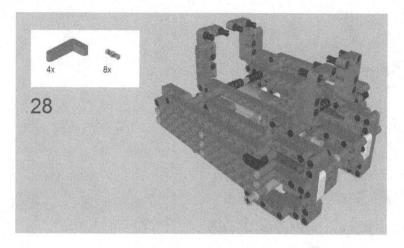

Figure 5-62. *Add the 5 × 3 Beams with the help of the Connector Pegs*

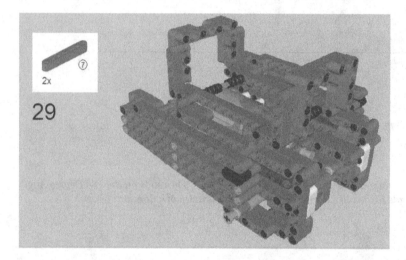

Figure 5-63. *This is the last step, joining with two 7M Beams*

If you attach the motor to a battery and turn on the power, you will see that it will work just like a scissor lift. It will hold a lot more weight than the rack-and-pinion model, and it can be adjusted for higher heights.

Summary

Considering that robots from science fiction have had extendable appendages for years, there is no reason why this can't be replicated in LEGO Technic.

There are several ways to do extensions on a LEGO Technic robot. The first way is to use a rack-and-pinion method, which involves a gear and a rack to extend out from where it normally wouldn't.

The other method also involves a rack-and-pinion method, but it uses a mechanism that can be seen on most scissor lifts.

Each method of extension can be motorized and bear a load. It is important to make certain your extension method isn't bearing too much weight or be overextended in any way.

The next chapter will discuss how to make a robot head, including very realistic facial features.

■ ■ ■

CHAPTER 6

The Robot Head

Okay, I realize that I am about to get geeky again, but I want you to think about all the robots you have seen in films. You will realize that most of them have heads that reveal something about their character.

In my earlier book I talked about how to create a wireframe for what you want to build. A wireframe involves planning a LEGO Technic project with a crude outline and then "thinking" in LEGO so you can make it work. In Chapter 2 of this book, you learned how to make certain shapes such as triangles, squares, and rectangles to help you create a robot body.

This chapter will show you how to create more curvaceous forms so you can create one of the most important parts of the robot—the head. If you think about robots in films, the ones with more curves to the face seem to look more human and therefore create a sense of familiarity. Think about C-3PO and R2-D2 from *Star Wars*, who have round and oval faces. Immediately these loyal droids stole our hearts in the first *Star Wars* Trilogy. In the prequel trilogy, the "bad guy" droids had flat and angular heads, as if they were manufactured for only military might the bad guys. Similarly, the Cylons from both versions of *Battlestar Galactica* were angular and triangular.

Granted, the humanoid exoskeletons from the *Terminator* films looked quite human, but their skull-like faces brought about a sense of fear. You have to think about what you want your robot face to convey. If you want people to love to hate your robot's face, then go with a villainous look.

I will also show you how to create a head for your robot and how you can "bring it to life." The three projects in this chapter will show you how to create a robot head complete with a moving mouth and even eyebrows.

You could make your robot's head smaller, I'll leave that up to you. The purpose of this is to show you how to make more curves, so you can get an idea of what to do when building your robot head. The robot head discussed in this chapter with Projects 6-1 to 6-3 will be slightly large in comparison to the robot body I showed you how to build in Chapter 3 of this book. The reason why I went with such a large robot head is so you can learn how to create a robot head with very human-like features and details. If you actually have enough pieces to create a LEGO Technic body to scale with this head, knock yourself out.

Before I get into building a robot head, I want to discuss how to replicate the curves that are part of the human head.

Circles, Curves, and Other Round Shapes

There is a motto LEGO builders use that I often see emblazoned on many T-shirts: Square Is Cool. I admit that it is much easier to work with right angles and even the triangles, which I discussed in Chapter 2.

It really isn't possible to make a circle using the current LEGO Technic parts, but there are ways around this. For example, if you link several #3 Angle Elements together with 2M Axles, you will get something that looks like a circle, as shown in Figure 6-1.

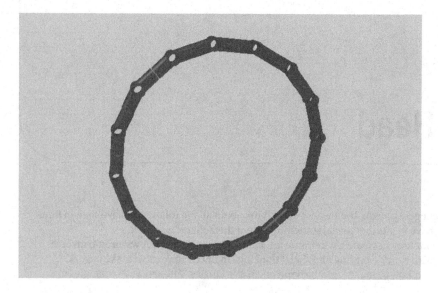

Figure 6-1. *What do you get when you cross sixteen 2M Axles with 16 #3 Angle Elements parts? A circle*

Figure 6-1 is actually a 16-sided figure, or hexadecagon, but I find that it does not fit well with certain figures. There is a distance of 2M between each round hole, provided they are close together, and you can bridge them with a 4M Lever and two Connector Peg/Cross Axles. I couldn't find a way to bridge any other hole with another hole. You can adjust the size of your hexadecagon by adjusting the size of the Axles, which increases the distance between the round holes.

If you use eight 2M Axles and eight #4 Angle elements you can produce a good looking octagon, as seen in Figure6-2.

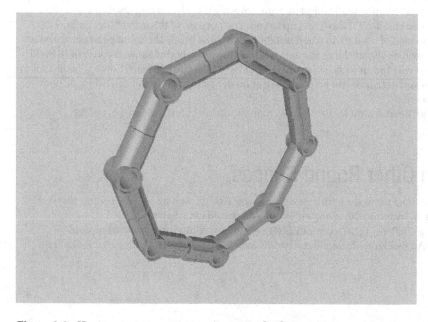

Figure 6-2. *How to create an octagon using #4 Angle Elements*

It is best to follow basic laws of geometry to create forms that will stand better, work better, and require less tweaking. For my earlier LEGO Technic book, I created an airplane model that used the Angular Beams to create the shape of an airplane fuselage that would slowly slope backward to the tail fin. After I had made it, and was pretty pleased with myself, I was prepared to use it in my book as a project.

However, when I tried to create this thing digitally, in LEGO Digital Designer (a digital LEGO building program that I discussed in Chapter 1), I discovered I couldn't build the same model in the digital world that I could in real life. I also discovered that if LEGO Digital Designer doesn't allow you to do a certain action with your LEGO pieces, it usually means there is something inherently wrong with the basic geometrical structure of your model.

You can imagine how frustrating it is when the program just wouldn't let me do something I could do in real life. I found out later that my model needed a lot of tweaking in order for to fit together and it may have not been a very good idea.

I found that it is possible to use LEGO Technic pieces to create just about any shape, but you want to make certain they all fit together without having to tweak them too much. It is even possible to create designs with curves such as circles by linking several pieces together. I have some projects that use curves, and this chapter will present one of them.

■ **Note** Before beginning this chapter's projects, refer to Appendix A for a complete list of required parts.

Project 6-1: Building the Top of the Head with Eyebrows

In Chapter 2, you learned how to create a robot body, and I included a section with squares, rectangles, triangles, and trapezoids. This first project is about making a construction with some serious curves.

In Figures 6-3 through 6-23 you will construct something to fit on top of the head, sort of like a scalp. In fact, you could actually construct this to fit over the top of your head. I actually wanted to do that originally, but I found that I couldn't fit the motors on for the eyebrows.

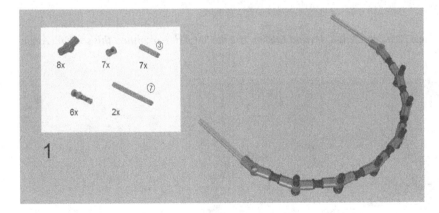

Figure 6-3. *With this step, you link together eight #3 Angle Elements with 3M Axles, with a Bush in the center of the 3M Axle. This creates the semicircular figure you see here, and you should place two 7M Axles on the ends along with Friction Snaps on the holes in the middle of the Angle Elements*

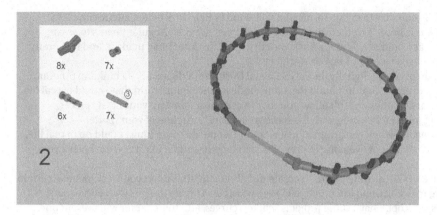

Figure 6-4. *Place the #3 Angle Elements onto the end of the 7M Axles and insert the Friction Snaps*

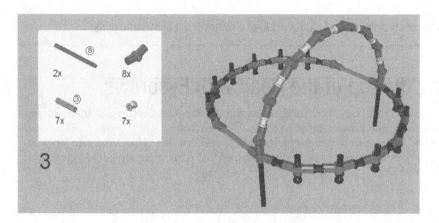

Figure 6-5. *Join another eight #3 Angle Elements with Axles and Bushes. Use the 8M Axles to connect this section's Angle Elements with the other, as shown*

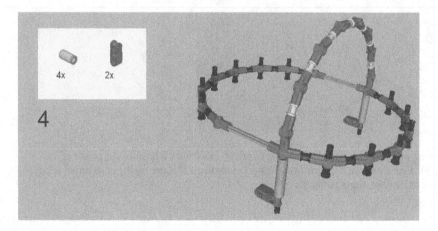

Figure 6-6. *Slide the Tubes onto the 8M Axles and then anchor them in place with the 3M Cross Blocks. There should be 1M of unused Axle at the bottom*

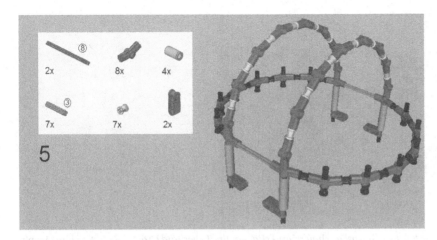

Figure 6-7. *This is essentially a repetition of Steps 3 and 4. Once again, you create another closed loop with the #3 Angle Element pieces, 3M Axles, and Bushes. Anchor it in place with the Tubes and 3M Cross Blocks. Note the position and direction of the 3M Cross Blocks*

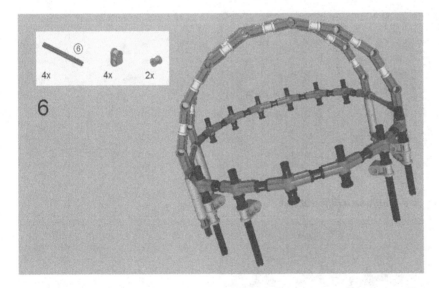

Figure 6-8. *Insert the 6M Axles into the Friction Snaps on one side of the construction. Slide the 90 Degree Cross Blocks and a Bush onto each side. The Axles are turned to the holes of the Cross Blocks to line up, which will be important in the next step*

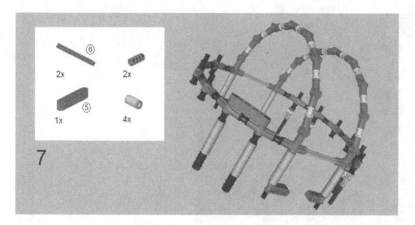

Figure 6-9. *Insert the 6M Axles in the Friction Snaps, as shown, and slide on the Tubes and Cross Axle Extensions. Don't forget to snap in the 5M Beam on top*

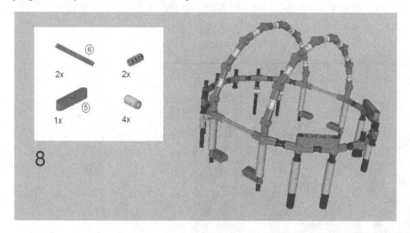

Figure 6-10. *This step is identical to Step 7, only on a different side*

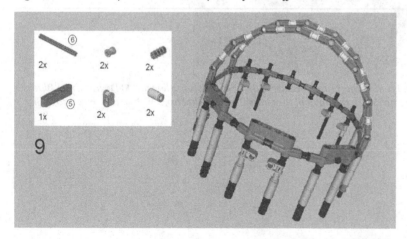

Figure 6-11. *Similar to Steps 7 and 8, this involves sliding on 6M Axles with 90 Degree Cross Blocks, Bushes, Tubes, and Cross-Axle Extensions in the center section, as well as the 5M Beam on top*

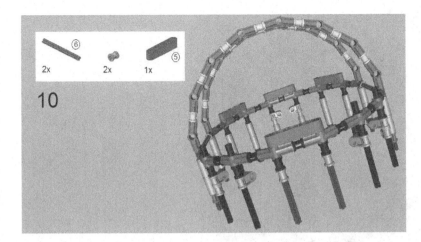

Figure 6-12. *Use the 5M Beam, 6M Axles, and Bushes on the other side of the construction*

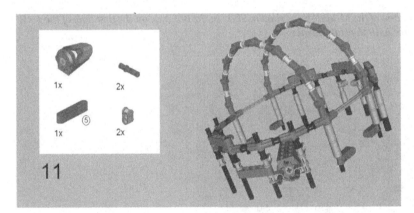

Figure 6-13. *Slide the 5M Beam onto the 6M Axles, as shown, and the 90 Degree Cross Blocks secure it in place. Snap in the 3M Connector Pegs until they are centered on the 90 Degree Cross Blocks, and then snap on the M-Motor*

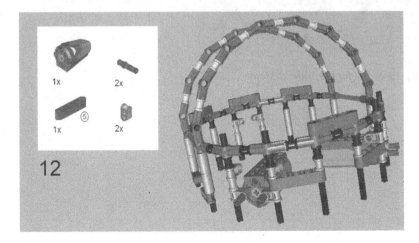

Figure 6-14. *This is essentially a repeat of Step 11, but on another side of the construction*

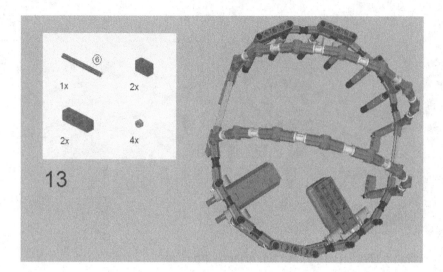

Figure 6-15. *Place the Technic bricks on one of the M-Motor's as shown. Slide in the 6M Axle on the first hole of the 1 × 2 Brick with two holes, and secure it in place with the Half Bushes*

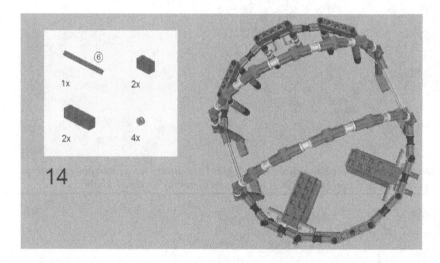

Figure 6-16. *This step is exactly like Step 13, except for the other M-Motor*

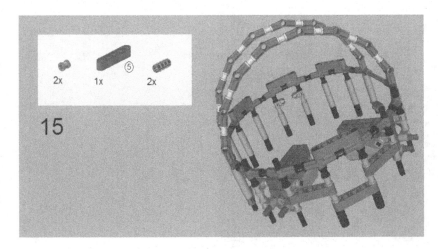

Figure 6-17. The Bushes, 5M Beam, and Cross-Axle Extensions go in between the M-Motors

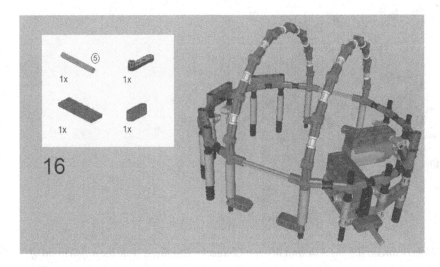

Figure 6-18. Use the 4M Lever with Notch on the M-Motor, and secure it into place with the 5M Axle and 3M Beam. Snap the 2 × 6 Plate into place on the Technic Bricks on the M-Motor

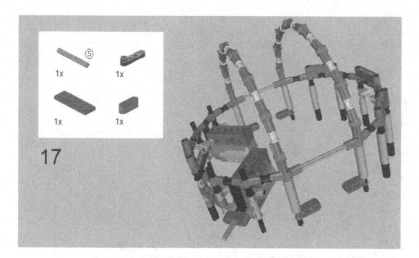

Figure 6-19. *This step is essentially identical to Step 17, except for the other M-Motor section*

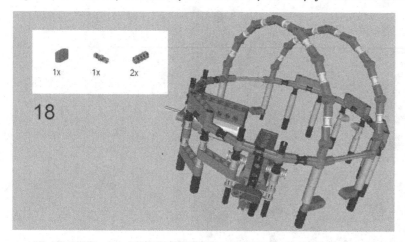

Figure 6-20. *Slide on a Cross and Hole Beam on the 5M Axle and insert a Connector Peg there. Note the application of the Cross-Axle Extensions on this side*

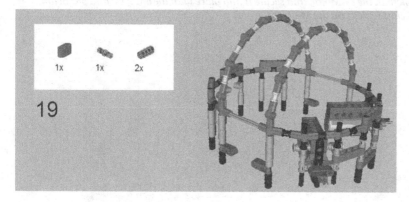

Figure 6-21. *This is identical to Step 18, except the focus is on the opposite side with the other M-Motor*

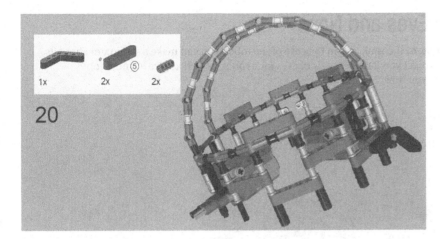

Figure 6-22. *Snap the 4 × 4 Angular Beam onto the Cross and Hole Beam. Slide on a 5M Beam and anchor it into place with the Cross-Axle Extension*

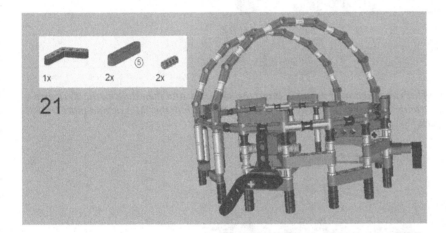

Figure 6-23. *The last step is identical to Step 20, with the 4 × 4 Angular Beam on the other M-Motor*

You will note that this model has eyebrows. Yes, I did say eyebrows. If you don't think that eyebrows are important, think about the movie *Short Circuit*. In the film, the robot known as Number Five is simply one in a quintet of robots that are built for blind military obedience. When Number Five is struck by lightning, it opens his mind up to being alive, or at least some form of sentience. Number Five doesn't have a lot of expression, but he does have two covers over his eyes that serve as artificial eyebrows.

Those tiny shifts on the eyes give Number Five (who later changes his name to Johnny Five) an illusion of life, and it was something that his robot cohorts failed to have. If you need another example, look at a dog and the shifting of his or her eyes. If you are a dog owner like myself, you know that this is the only window you have to figure out what yourdo gisf eeling.

In case you haven't yet realized it, the motors are used to control the expression of the eyebrows. You will need to connect the IR-RX and one of the remote controls to make certain it works. I let you decide where to put those, but for now, let's add some eyes and nose to your robot head.

Project 6-2: Adding Eyes and Nose

They say that the eyes are the window to the soul, and in the case of the robot, you can make a statement about its character through its eyes. In this project you'll make a pair of eyes for a robot, as well as give it a nose job. Figures 6-24 through 6-44 will show you how.

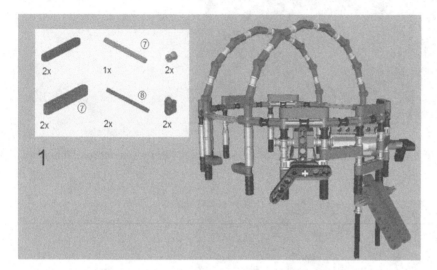

Project 6-24. *This is the beginning of the construction of the nose. Insert the 8M Axles, and then slide on the 90 Degree Cross Blocks, anchoring them in place with the Bushes. Join together the Cross Blocks with the 7M Axle and place the 6M Half Beams and 7M Beams in between*

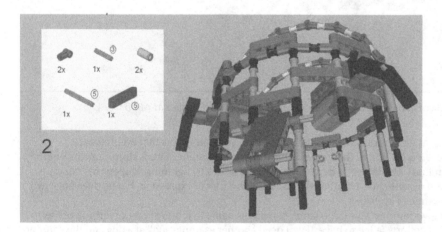

Project 6-25. *Slide in the 5M Axle with one end of the 5M Beam in place. Slide in the Tubes, and then the Zero Degree Elements are attached to the 5M Beam*

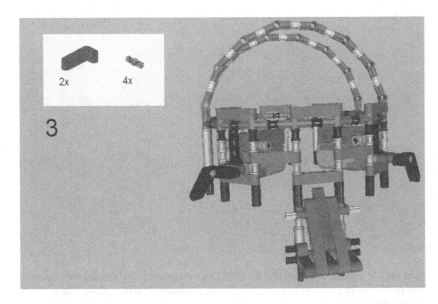

Figure 6-26. *Insert the 4 × 2 Angular Beams on the 5M Axle, through the cross hole. Note the location of the four Connector Pegs*

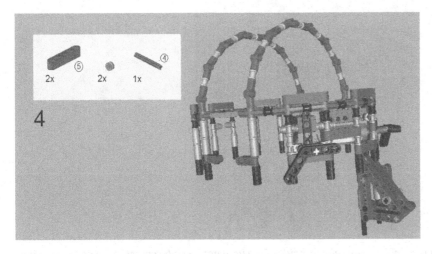

Figure 6-27. *Snap the 5M Beams onto the side of the nose to secure the 4 × 2 Beams in place, and slide the 4M Axle into place on the "slant" and secure it with the Half Bushes*

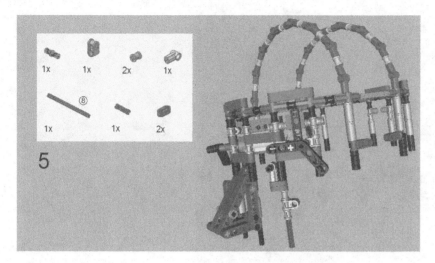

Figure 6-28. *Insert the 8M Axle and then the 2M Levers. Use the 2M Axle, Zero Degree Element, and Connector Peg on the other cross hole of the 2M Lever. Slide on the Bushes and 90 Degree Cross Block, as shown. (I will explain why the 90 Degree Cross Block is put here at the end of this project.)*

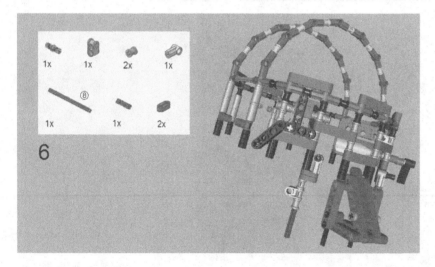

Figure 6-29. *This step is identical to Step 5, but on the other side of the nose note the placement of the parts. Note the placement of the parts*

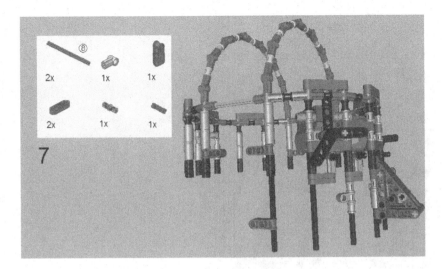

Figure 6-30. *Like Steps 5 and 6, you need to slide Levers onto an 8M Axle. Use a 2M Axle, Zero Degree Element, and Connector Peg in a similar fashion. Insert the 8M Axle and slide onto a 3M Cross Block*

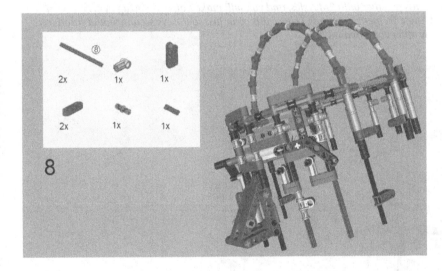

Figure 6-31. *This step is just like Step 7, but on the opposite side*

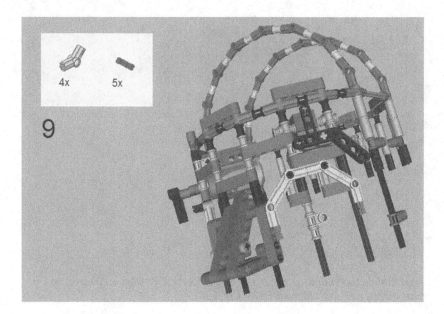

Figure 6-32. *Join four #4 Angle Elements together with 2M Axles, and you will make half of an octagon to create an "eye." (It may be difficult to find a #4 Angle Element in yellow; I chose this color just so the eye section would stand out on this illustration. Feel free to use any other color available.)*

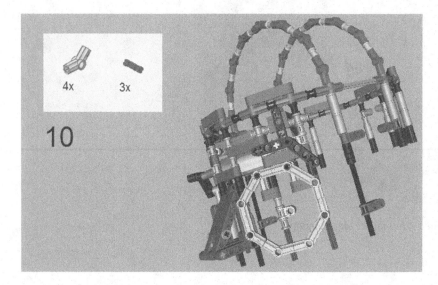

Figure 6-33. *Join together the four #4 Angle Elements to form another half of an octagon with the 2M Axles. Slide it on to complete the "eye"*

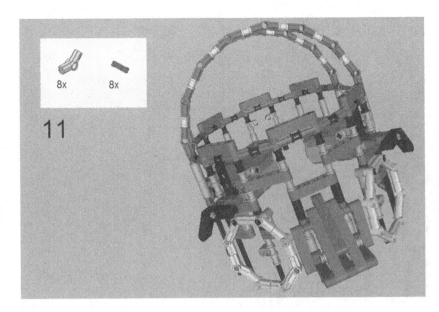

Figure 6-34. *Like Steps 9 and 10, it is time to create another octagon to make the other "eye" and then snap it into place on the Connector Pegs*

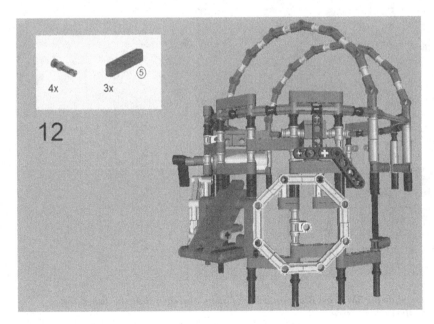

Figure 6-35. *Slide on three 5M Beams on the bottom, as shown. Anchor them into place with the Friction Snaps*

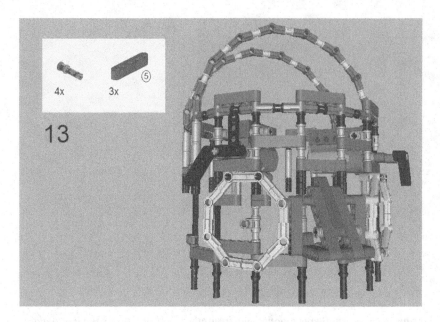

Figure 6-36. *This is another step that is similar to Step 12*

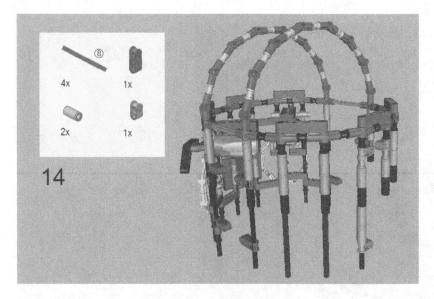

Figure 6-37. *Insert 8M Axles as shown. Slide the 3M Cross Block onto one of them, and then slide the Tubes and 90 Degree Cross Block onto the Axle on the opposite side*

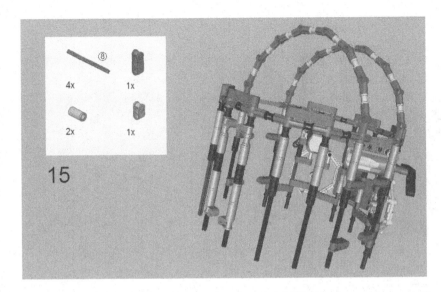

Figure 6-38. *This step is like Step 14, but on the opposite side of the creation*

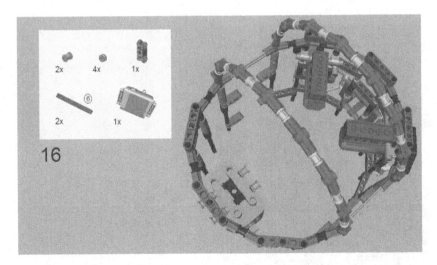

Figure 6-39. *On top of the creation, slide in the 6M Axles through some 90 Degree Cross Blocks. Insert the Half Bushes, and slide on the battery with the Double Cross Block in the middle. Secure with Bushes (the battery will be more secured in the next step)*

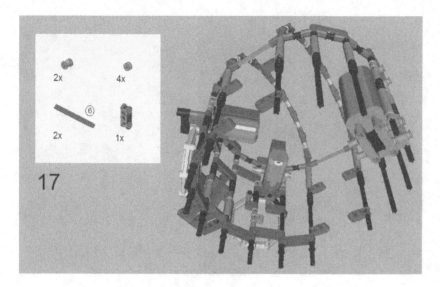

Figure 6-40. *This is similar to Step 16, except that the Bushes, Half Bushes, 6M Axles, and Double Cross Block intersect with the battery box at the bottom instead of the top*

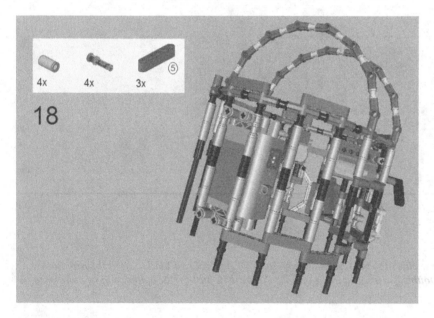

Figure 6-41. *Slide on the Tubes, and then the three 5M Beams. Anchor everything in place with the four Friction Snaps*

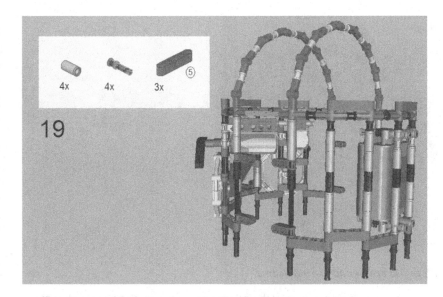

Figure 6-42. *Again, another step where you imitate the previous one, but on the opposite side*

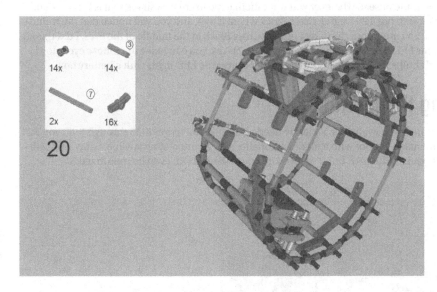

Figure 6-43. *This step is very similar to Steps 1 and 2 in Project 6-1, and it serves as a foundation for the eyes and nose section as it is snapped into place on the Friction Snaps*

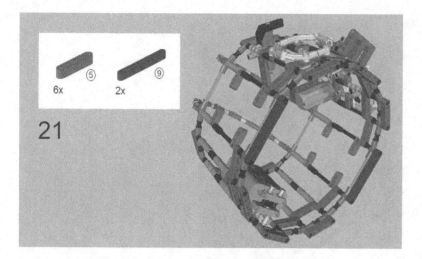

Figure 6-44. *Cap off the bottom with six 5M Beams and two 9M Beams*

I will let you decide what kind of character the eyes and nose of this robot convey. Perhaps you want to change them, sof eelf reetodo this .I a ma ssumingn oon eh astr ademarkedthe r obothe ad.

Of course, you can alter the eyes or the nose to whatever you want them to be to create a shape that is less human. I mentioned in Step 5 that I had a reason to put a 90 Degree Cross Block where it was, and you will notice that the round hole is in the center of the eye. In case you are wondering why I put the Cross Block in the middle of the eyes, I did this so I can insert the LEGO Power Functions LED lights (see Chapter 3, Figure 3-7). Yes, you can make the robot's eyes glow! All that is required is to insert the LED bulbs through the round holes and power the LED lights with a battery box.

Project 6-3: Creating the Mouth

Now that your robot has eyes, nose, and even eyebrows, it clearly needs a mouth. In Figures 6-45 through 6-65 you will bring all of these things together and make a robot with a lot of personality and an interesting jawline. Sadly, you will have to work the robot like a ventriloquist's dummy, but it really works out well and will even be motorized.

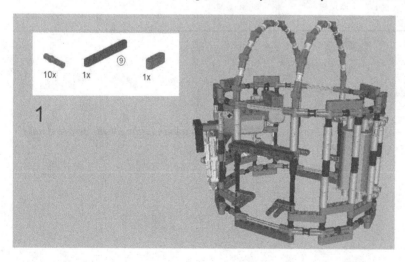

Figure 6-45. *Insert the 3M Connector Pegs through the 3M Cross Blocks. Insert a 9M Beam on the top row and put the 3M Beam on the other side*

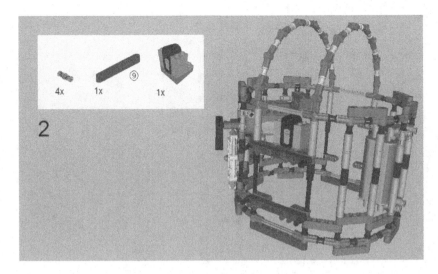

Figure 6-46. *Insert another 9M Beam on the bottom row of the 3M Connector Pegs, then put a Connector Peg on each of the ends of the 9M Beams. Snap the IR-RX into place on the top, as shown*

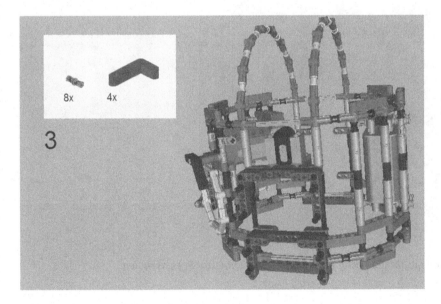

Figure 6-47. *Snap the 5 × 3 Angular Beams onto the 9M Beams and then insert the Connector Pegs, as shown*

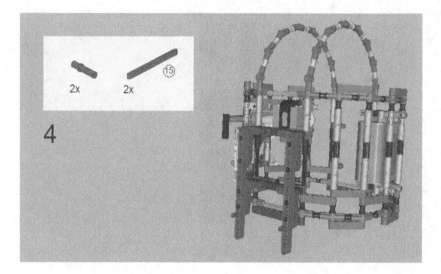

Figure 6-48. *Snap on the 15M Beams and the 3M Connector Pegs, as shown*

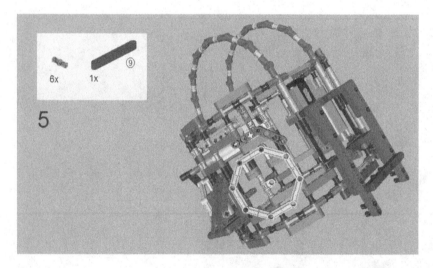

Figure 6-49. *Insert the 9M Beam on the 3M Connector Pegs. Snap on the Connector Pegs, as shown*

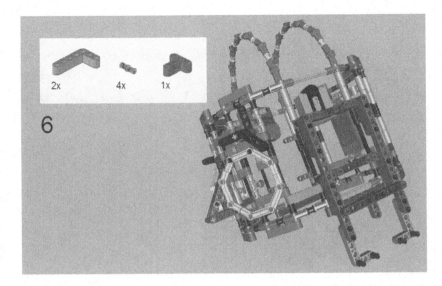

Figure 6-50. *Center the 3 × 3 or T-Beam on the 9M Beam. Place the 5 × 3 Beams on the bottom and the Connector Pegs on them*

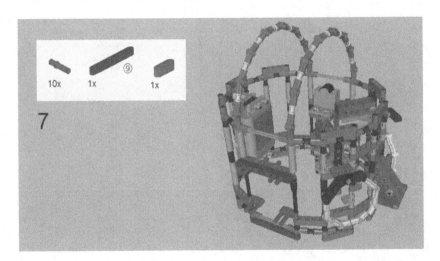

Figure 6-51. *Turn the creation around. Insert the 3M Connector Pegs through the 3M Cross Blocks. Insert a 9M Beam on the top row and put the 3M Beam on the other side*

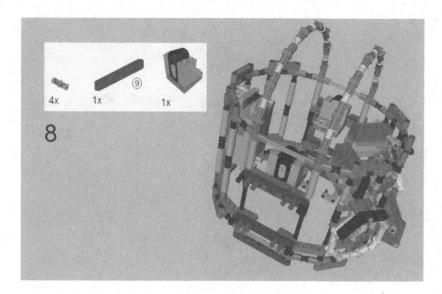

Figure 6-52. *Insert another 9M Beam on the bottom row of the 3M Connector Pegs, then put a Connector Peg on each of the ends of the 9M Beams. Snap the IR-RX into place on the top, as shown*

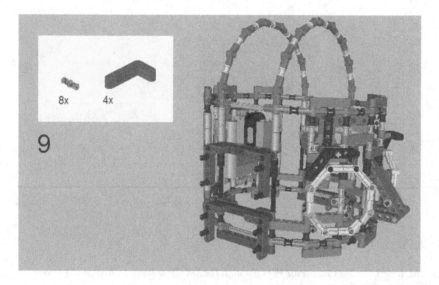

Figure 6-53. *Snap the 5 × 3 Angular Beams onto the 9M Beams and then insert the Connector Pegs, as shown*

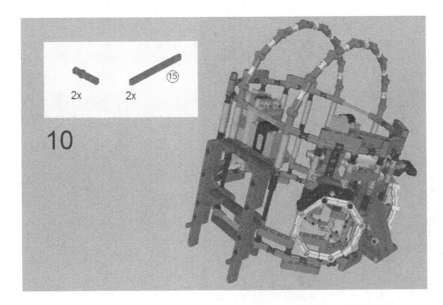

Figure 6-54. *Snap on the 15M Beams and the 3M Connector Pegs, as shown*

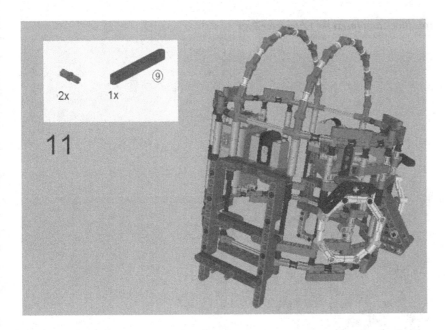

Figure 6-55. *Snap on the 9M Beams and the Connector Peg/Cross Axles, as shown*

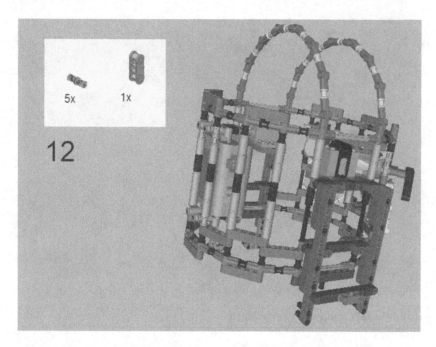

Figure 6-56. *Slide on the Double Cross Block and insert the Connector Pegs*

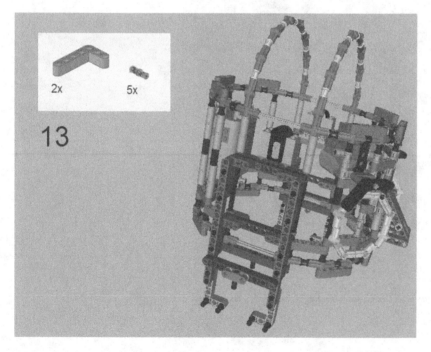

Figure 6-57. *Insert the 5 × 3 Angular Beams on the bottom of the structure, with the Connector Pegs on that*

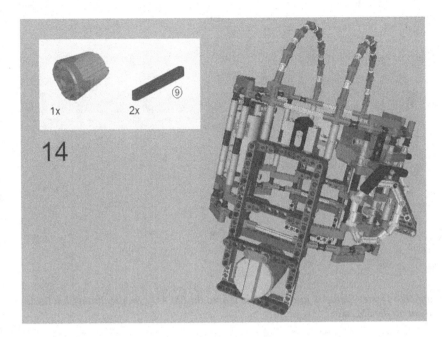

Figure 6-58. *Connect the XL-Motor to the Connector Pegs that hang vertically, then snap the 9M onto the horizonal Connector Pegs below it*

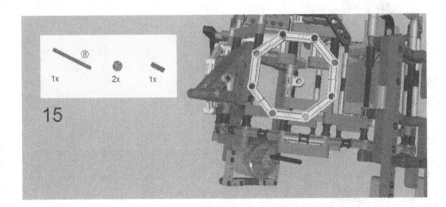

Figure 6-59. *Insert the 2M Axle in the center of the XL-Motor and place an 8-tooth Gear on that. Place an 8M Axle on the round hole above that and place an 8-tooth Gear on that*

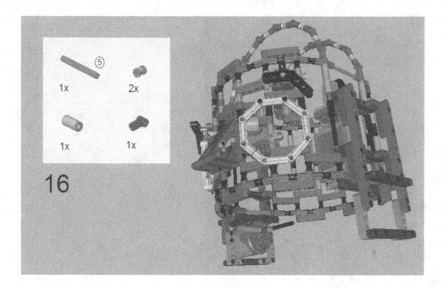

Figure 6-60. *Insert the 5M Axle on the Zero Degree Element, and then slide it onto the 8M Axle. Secure that with a Bush on the 8M Axle and put a Tube and Bush on the 5M Axle*

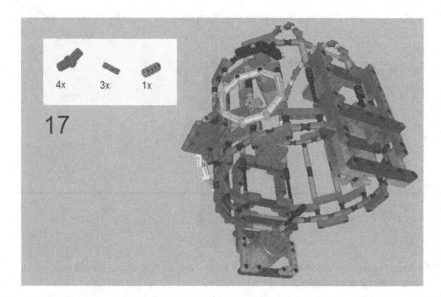

Figure 6-61. *Connect four #3 Angle Elements together with the 2M Axles, and place a Cross-Axle Extension on the end of the 8M Axle*

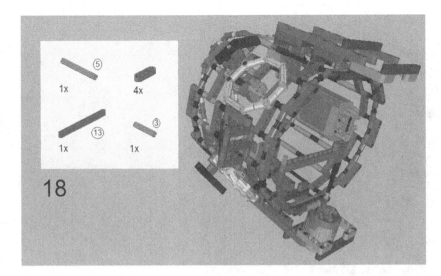

Figure 6-62. *Slide a 5M Axle on the Cross-Axle Extension. Slide on two 3M Levers and put a 3M Axle on the other end. Put on a 13M Beam so it intersects with the 3M Axle, and then put the other two 3M Levers on*

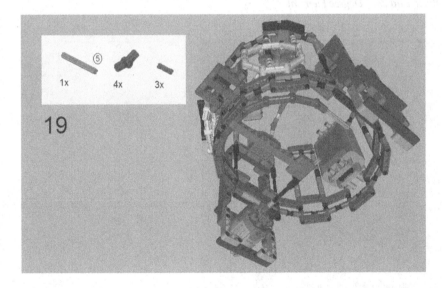

Figure 6-63. *Attach the rest of the #3 Angle Elements on with the 2M Axles. Place the 5M Axle on the last #3 Angle Element*

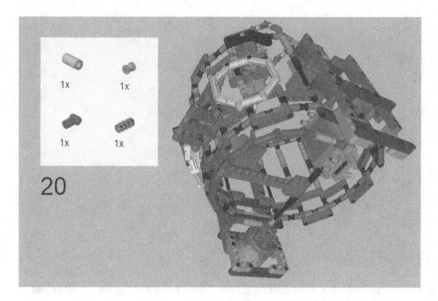

Figure 6-64. *Place a Cross-Axle Extension on the end of the 5M Axle after the 3M Levers. On the other 5M Axle after the #3 Angle Element, place the Tube, Bush, and Zero Degree Element*

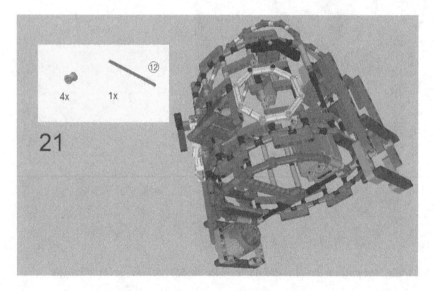

Figure 6-65. *Insert the 12M Axle and anchor it into place with the four Bushes*

Bringing the Robot Head to Life

You may notice that the eyebrows of Project 6-1 have the 4M Lever with Notch. The reason why that part is there is so the eyebrows only shift a little and don't go around in circles. This is something that often has to be put on creations so that the motor doesn't carry something too far. In the same manner, you should put some sort of catch on the mouth so the jaw doesn't go crazy.

While I am on the subject of motorizing things, I should mention how to connect the motors and IR-RX. You should connect the two M-Motors that control the eyebrows to one IR-RX and then take control of them with a remote control.

As for the other IR-RX, you should definitely connect the XL-Motor to it. Since you have another port, you might as well make a neck. Fortunately, you have already learned the technique of how to make a robot's neck, as this is the same mechanism you used for the robot wrist in Chapter 4, Project 4-2.

At this point, I am going to assume that you have an ability to "think in LEGO" enough to create a smaller version of this large head. With this particular swivel section, you should be able to mount any type of head and swivel it around like a neck. In fact, you should be able to make the head go around 360 degrees, or 720 degrees, better than anyo wl.

Summary

When it comes to making a robot head, it is more about creating a head that shows character and personality.

By creating a robot with some working eyebrows, you can make a robot that can express both anger and surprise. The eyes also convey something that can almost be described as human.

It is also possible to make a mouth that moves if you want to give the robot the illusion of speech.

Now that we have finished a robot's body, arms, and head, the final chapter will show how to make a robot walk witht woo rm orel egs.

CHAPTER 7

Enabling a LEGO Technic Robot to Walk

This is the last chapter in this book, and you can probably guess without even looking at the title what it will be about. I have thoroughly covered how to construct just about every shape for a robot body, how to give it arms, even constructing a head. I covered how to give it motion, but only with wheels. Wheels are great if you want to build an R2-D2 type of robot, but if you want it to walk like C-3PO, then you will have to give it a pair of legs.

This chapter will show you how to make a LEGO robot walk. It will cover how to make a pair of legs, but not necessarily how to structure the shape of them. When I played with LEGOs as a child, I attempted to make walking models. The difficulty was not creating the legs, but rather in building a mechanism that would allow them to walk. I had a method that used 2 × 2 Turntable pieces in a way so the legs would shift back and forth, but would never lift off the ground.

Here, you'll learn how to make legs that can lift off the ground, and, with the application of some gears, it will move. It is not the only way to make a robot walk, and you can find many LEGO creations on the Internet that walk in ways that I couldn't even conceive. Just go to a video-sharing web site like YouTube and type in LEGO walker or similar keywords and you can view videos on how others have gotten them to walk.

The trick of creating a LEGO Technic walking robot is to create something that is motorized but can keep the feet stepping without it tipping over. Project 7-1 will show you how to create a motorized section that will power the legs. Project 7-2 will show you how to create a pair of legs and hook them up to Project 7-1.

Note   Before beginning this chapter's projects, refer to Appendix A for a complete list of required parts.

Project 7-1: A Two-Legged Walker Motor

This walking robot project is kind of an accumulation of a dream for me. As I explained above, I have been trying for some time to enable a LEGO creation the ability to walk. The best I could do was a method where the feet would shuffle forward one at a time, but this was only an illusion of walking. Most of my walking creations were just wheels that had legs that "appeared" to be pedaling them along, as if it were a LEGO person riding a bicycle. I never figured out how to make a two-legged creation lift even one leg without the entire thing tipping over.

Until now. The key is making certain that the robot shifts its weight and still keeps standing, just like a real person walks. The secret is in the design of the legs, but first you need something that will put the legs into motion. Project 7-1 will create the motorized section to power the legs for the robot. Follow the 37 steps presented in Figures 7-1 through 7-37f ort hef irstp arto ft hisb uild.

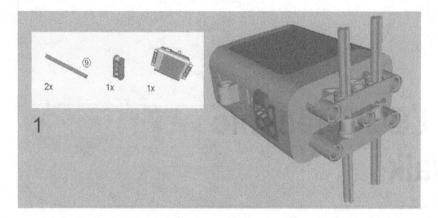

Figure 7-1. *Insert the two 9M Axles into the battery pack, with 3M of Axle on each end. The Double Cross Block holds them in place, and they will be made more secure in later steps*

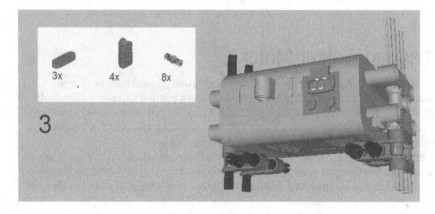

Figure 7-2. *This step is a lot like Step 1, but it uses 8M Axles instead of 9M. They need to be centered in, which means about 2 1/2 M on each side*

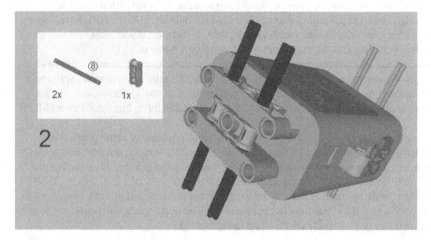

Figure 7-3. *Insert a 3M Lever on each side, and then slide a 3M Cross Block on each of the Axles and place Connector Pegs in all the round holes. Insert another lever at the bottom on one the side with the 9M Axles*

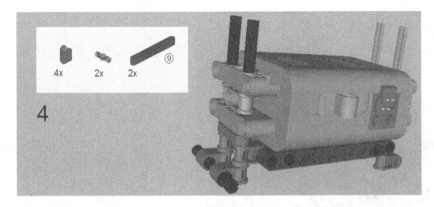

Figure 7-4. *Insert 9M Beams on each side of the construction, which will hold the 3M Cross Blocks from Step 3 in place. Slide on 90 Degree Cross Blocks on each side and insert the Connector Peg on the side, as shown*

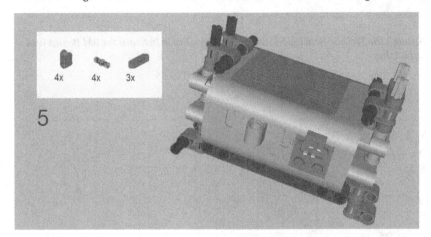

Figure 7-5. *This step is similar to Step 3, but instead of using 3M Cross Blocks, it uses 90 Degree Cross Blocks. Insert the Connector Pegs on each of the 90 Degree Cross Blocks*

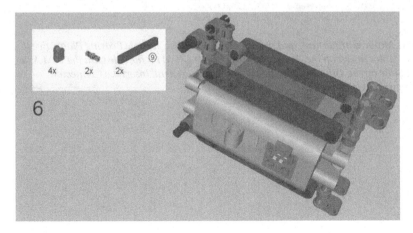

Figure 7-6. *Insert the 9M Beams onto the Connector Pegs from Step 5. Slide the 90 Degree Cross Blocks onto the top of the Axles and insert the additional Connector Pegs*

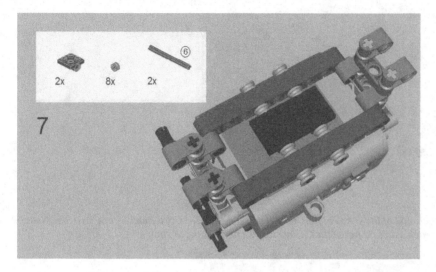

Figure 7-7. *Slide the two 6M Axles through the Technic Bearing Plates and Half Bushes in between the 9M Beams and then secure the 6M Axles with two Half Bushes*

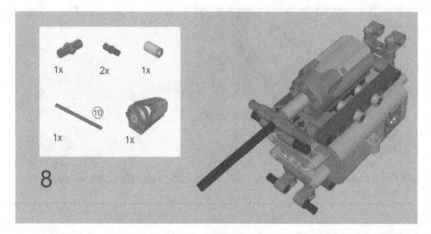

Figure 7-8. *The exact placement of the M-Motor is important, and the hollow studs on the Technic Bearing Plates give it a lot room to move. Be certain that the M-Motor hangs off the Technic Bearing Plates 1 ½ M on the side with the 10M Axle and a ½ M on the side opposite the Axle. Slide on the Tube and the 180 Degree Angle Element. Insert the Connector Peg/Cross Axle on each side of the 180 Degree Angle Element*

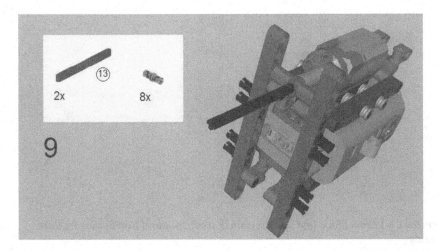

Figure 7-9. *Snap the 13M Beam onto each side of the Connector Peg/Cross Axles and then insert four Connector Pegs, as shown*

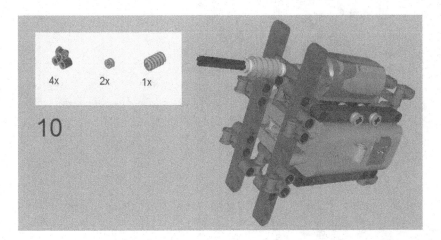

Figure 7-10. *Snap the 2 × 1 Cross Blocks onto the Connector Pegs from Step 9. On the 10M Axle, slide on the Half Bush, Worm Gear, and another Half Bush*

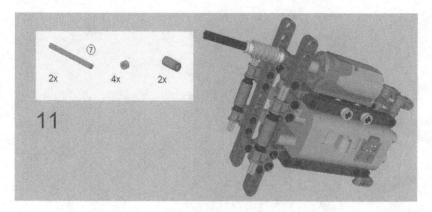

Figure 7-11. *Slide the 7M Axles onto the 2 × 1 Cross Block, and make certain that the Tube and Half Bushes are there. There should be 1M of Axle on each side*

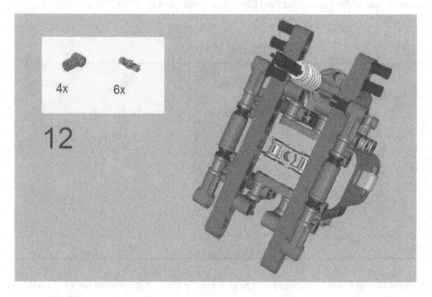

Figure 7-12. *Place the Zero Degree Elements on the ends of the 7M Axles from Step 11. Snap the Connector Pegs onto the 13M Beams and two of the Zero Degree Angle Elements*

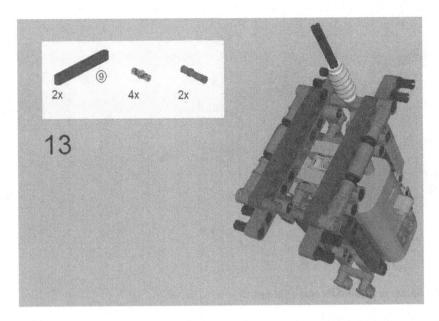

Figure 7-13. *Insert the four Connector Pegs at the bottom and use the 3M Connector Pegs to secure the 9M Beams*

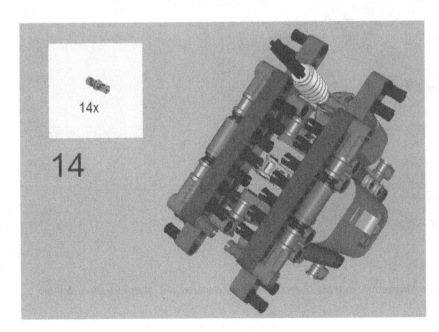

Figure 7-14. *Place all of the Connector Pegs in the center of the 9M Beams*

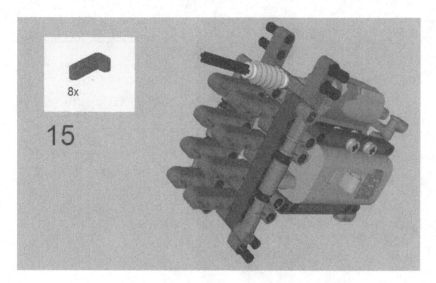

Figure 7-15. *Snap the 4 × 2 Angular Beams onto the sides of the 9M Beams*

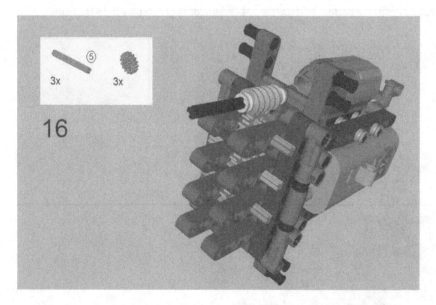

Figure 7-16. *Slide in the 5M Axles, making certain that the 16-tooth Gears are centered and meshing with each other and the Worm Gear*

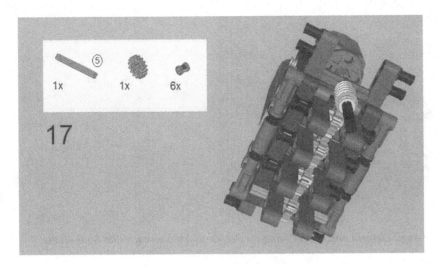

Figure 7-17. *Insert one more 5M Axle and 16-tooth Gear at the bottom. Secure the other three 5M Beams with Bushes on each end*

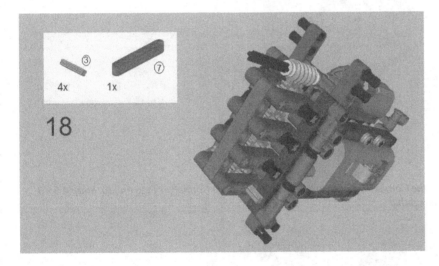

Figure 7-18. *Insert the 7M Beam between the 4 × 2 Beams and then secure it into place with the four 3M Axles*

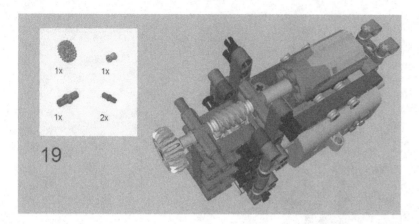

Figure 7-19. *Slide on the 180 Degree Angle Element and insert Connector Peg/Cross Axles on each side. Slide on the Bush and Z20 Gear*

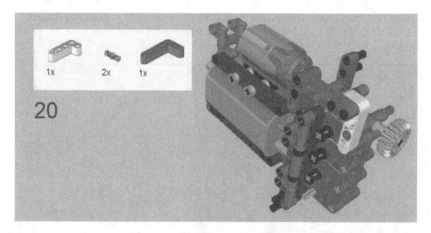

Figure 7-20. *Insert a 4 × 2 Beam on the Connector Peg/Cross Axle and place two Connector Pegs on top. Snap a 5 × 3 Angular Beam in place to hold it all together*

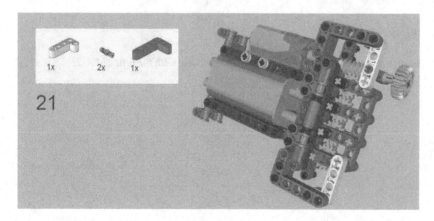

Figure 7-21. *This is just like Step 20, except it is on the bottom of the creation. There is no Connector Peg/Cross Axle here*

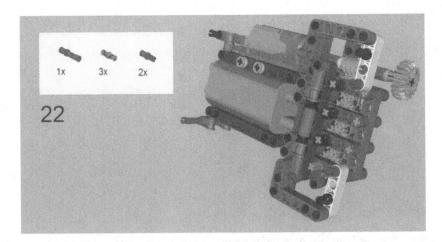

Figure 7-22. *Insert the 3M Connector Peg on the far left side of the creation and place a Connector Peg above that. Insert the Connector Peg/Cross Axles and Connector Pegs on the 4 × 2 Angular Beams*

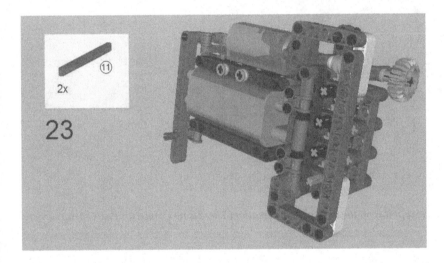

Figure 7-23. *Snap 11M Beams onto each side*

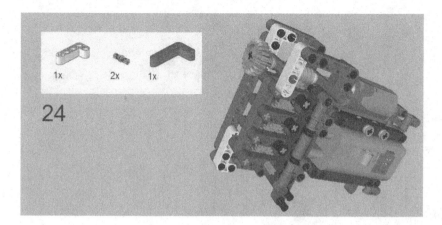

Figure 7-24. *Insert a 4 × 2 Beam on the Connector Peg/Cross Axle and place two Connector Pegs on top. Snap a 5 × 3 Angular Beam in place to hold it all together*

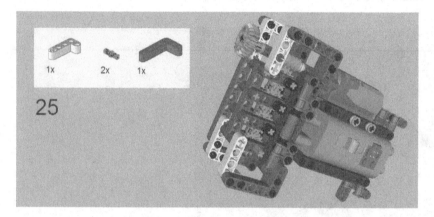

Figure 7-25. *This is just like Step 24, except it is on the bottom of the creation. There is no Connector Peg/Cross Axle here*

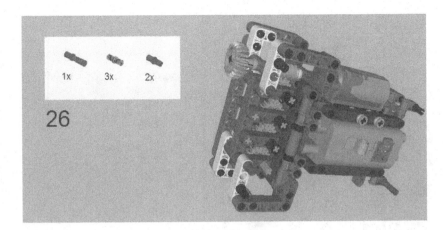

Figure 7-26. *Insert the 3M Connector Peg on the far right side of the creation and place a Connector Peg above that. Insert the Connector Peg/Cross-Axles and Connector Pegs onto the 4 × 2 Angular Beams*

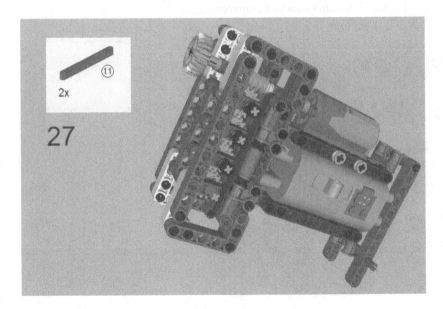

Figure 7-27. *Snap 11M Beams onto each side*

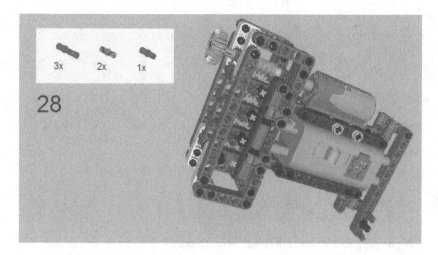

Figure 7-28. *Insert three 3M Connector Pegs with the ring side down, with a Connector Peg below that. On the other side of the creation, insert the Connector Peg/Cross Axle and another Connector Peg*

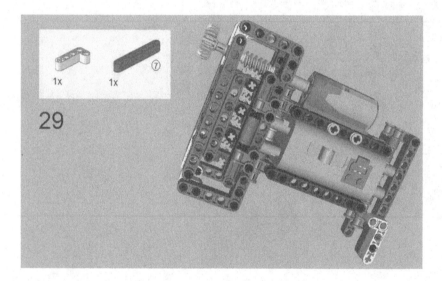

Figure 7-29. *Snap the 7M Beam onto the three 3M Connector Pegs from Step 28. Snap on the 4 × 2 Angular Beam from the opposite side*

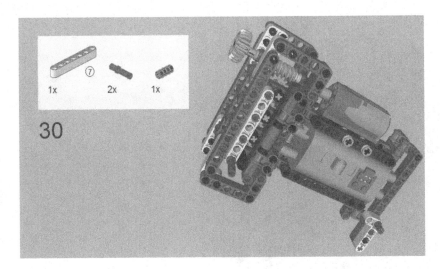

Figure 7-30. *Snap a 7M Beam atop the 7M Beam from Step 29. Snap on the 3M Connector Pegs. Slide the Cross-Axle Extension on one end of the 5M Axle*

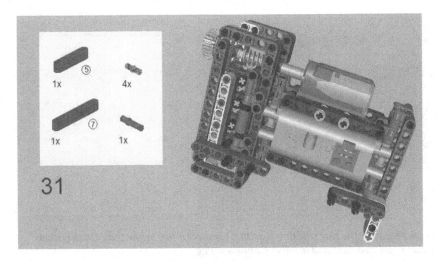

Figure 7-31. *Snap a 3M Connector Peg onto the 4 × 2 Angular Beam and then snap on the 5M Beam. Snap the 7M Beam onto the other side, and then snap on the Connector Pegs*

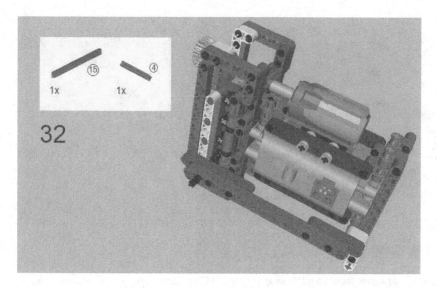

Figure 7-32. *Snap on a 15M Beam and slide in a 4M Axle on the Cross Axle Extension from Step 30*

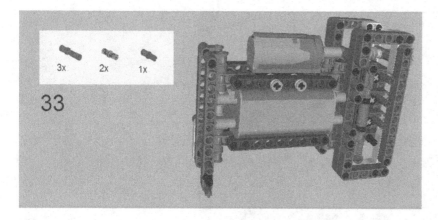

Figure 7-33. *Insert three 3M Connector Pegs with the ring side down, with a Connector Peg below that. On the other side of the creation, insert the Connector Peg/Cross Axle and another Connector Peg*

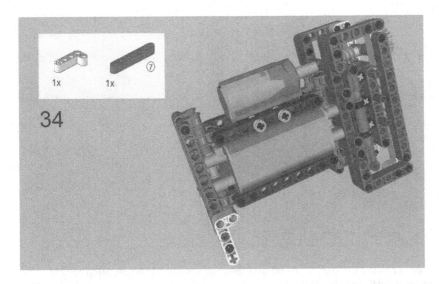

Figure 7-34. *Snap the 7M Beam onto the three 3M Connector Pegs from Step 33. Snap on the 4 × 2 Angular Beam from the opposite side*

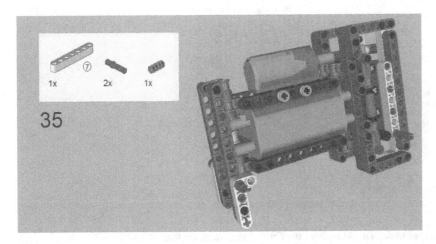

Figure 7-35. *Snap a 7M Beam atop the 7M Beam from Step 34. Snap on the 3M Connector Pegs. Slide the Cross Axle Extension onto one end of the 5M Axle*

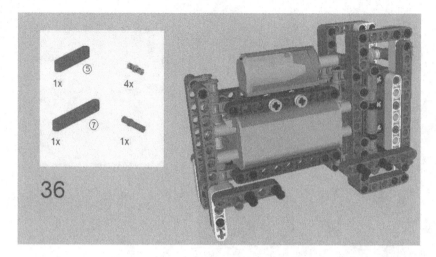

Figure 7-36. *Snap a 3M Connector Peg onto the 4 × 2 Angular Beam and then snap on the 5M Beam. Snap the 7M Beam onto the other side and then snap on the Connector Pegs*

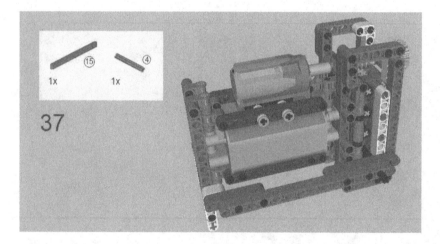

Figure 7-37. *Snap on a 15M Beam and slide in a 4M Axle on the Cross Axle Extension from Step 35*

Project 7-2: The Legs of the Two-Legged Robot Walker

Of course all that we have really created in Project 7-1 is a way to connect a battery and a motor together so we can have both power and action. We could now put wheels on the 4M Axles, and yes, it would move. However, we want to make it walk, so we are going to need legs.

This is of course not the only way of making legs, no more than Project 7-1 is the only way to motorize them. You will note that the legs have certain "joints" that allow for free spinning, and yet these spinning places are always locked in the same place. This allows a leg to move in a circular fashion, and the other to move similarly. This combined motion will allow a walking creation to move across the table, and I highly recommend a hard surface like a tabletop and not a carpet.

Project 7-2 will show you how to make a pair of legs and then hook them up to the motorized section of the robot. The 18 steps presented in Figures 7-38 through 7-55 will guide you in completing the leg mechanisms and placing themo nP roject7 -1.

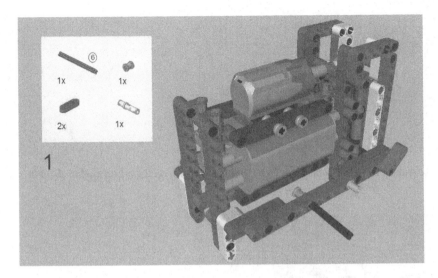

Figure 7-38. *Snap in a 3M Connector Peg along with a 6M Axle with Bush. Insert two 3M Levers on the end of the 4M Axle. Please note the angle of the 3M Levers, as this will be important when the other leg is constructed on the opposite side*

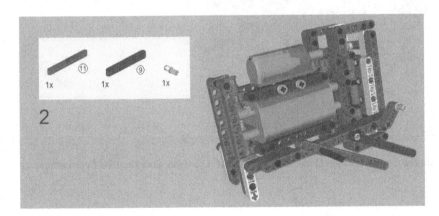

Figure 7-39. *Slide on an 11M Beam and a 9M Beam and a Connector Peg/Cross Axle at the end of the 3M Lever. Don't worry about the angles of the Beams, as the next step will secure them in place*

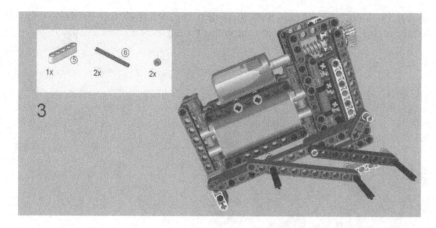

Figure 7-40. *Slide a 6M Axle onto the ends of the two Beams from Step 2. Put a 5M Beam on them and anchor them in place with the Half Bushes*

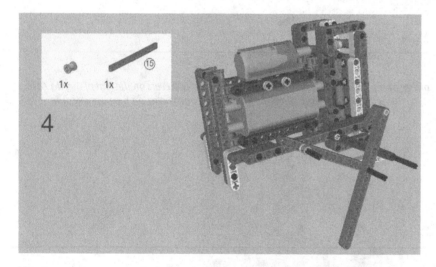

Figure 7-41. *Insert a 15M Beam onto the Connector Peg/Cross Axle and the 6M Axle. Note how the 15M Beam intersects with the 6M Axle*

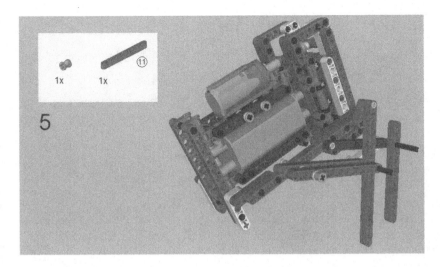

Figure 7-42. *Slide a 11M Beam onto the 6M Axle and get a Bush to anchor it into place*

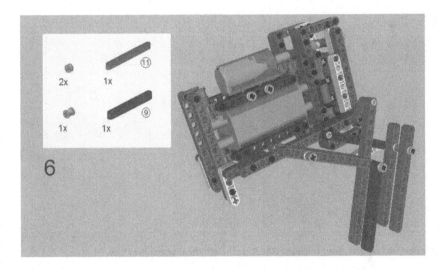

Figure 7-43. *Slide on the 9M and 11M Beams and anchor them into place with the Bush and two Half Bushes*

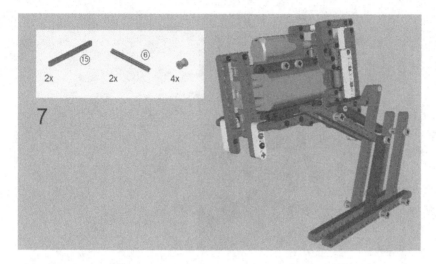

Figure 7-44. *Slide the 6M Axles through the bottom and the 15M Beams. Anchor them into place with Bushes*

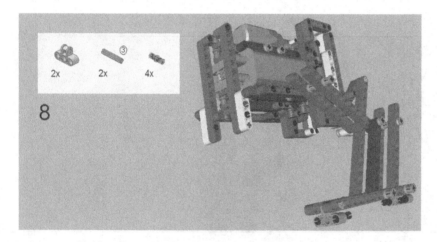

Figure 7-45. *Insert the 2 × 3 Cross Blocks and secure them with the 3M Axle. Insert the Connector Pegs*

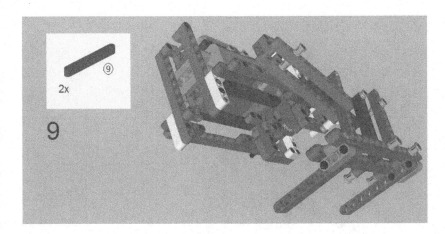

Figure 7-46. *Snap the 9M Beams into place at the bottom*

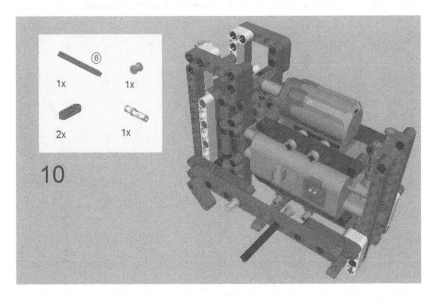

Figure 7-47. *Now it is time to work on the other side of the creation. Insert two 3M Levers on the end of the 4M Axle. Snap in a 3M Connector Peg along with a 6M Axle with Bush*

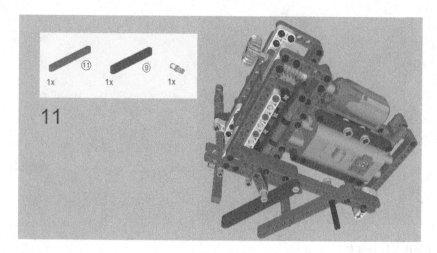

Figure 7-48. *Slide on an 11M Beam and a 9M Beam and add a Connector Peg/Cross Axle at the end of the 3M Lever*

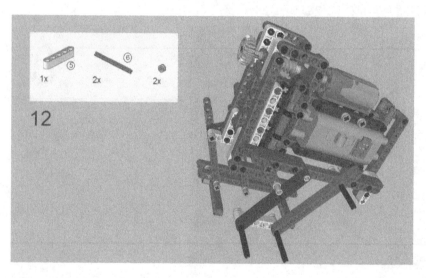

Figure 7-49. *Slide a 6M Axle onto the ends of the two Beams from Step 11. Put a 5M Beam on them and anchor them in place with the Half Bushes*

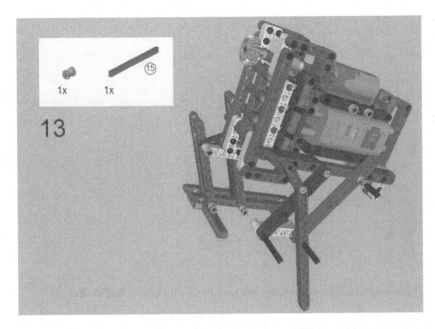

Figure 7-50. *Insert a 15M Beam on the Connector Peg/Cross Axle and the 6M Axle*

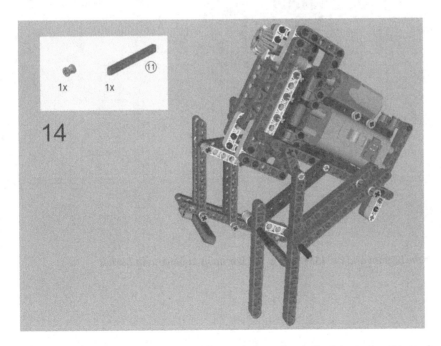

Figure 7-51. *Slide a 11M Beam onto the 6M Axle, and set a Bush to anchor it into place*

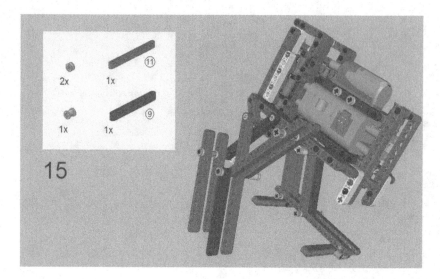

Figure 7-52. *Slide on the 9M and 11M Beams and anchor them into place with the Bush and two Half Bushes*

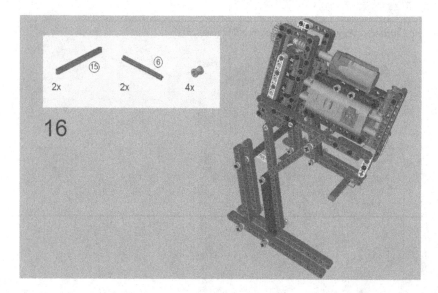

Figure 7-53. *Slide 6M Axles through the bottom and the 15M Beams. Anchor them in place with Bushes*

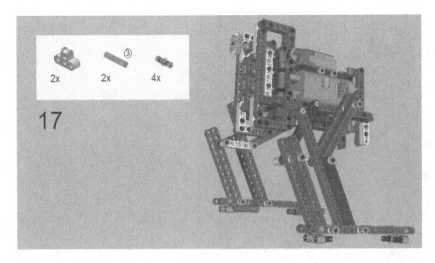

Figure 7-54. *Insert the 2 × 3 Cross Blocks and secure them with the 3M Axle. Insert the Connector Pegs*

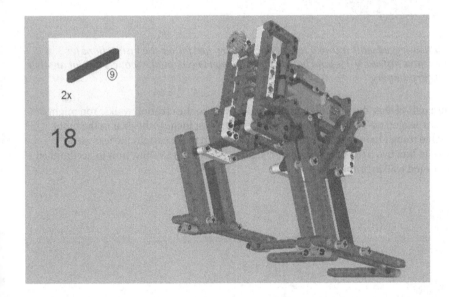

Figure 7-55. *Snap the 9M Beams into place*

Creating a Walker with More Than Two Legs

Now that you have created a walker with two legs, you might want to take it a step further and create one with four legs. I found that it really isn't too difficult to simply create another two-legged walker and link it to the first set. You will find that you won't need another battery for that, so I recommend instead of using the battery box in your second two-legged walker that you replace it with four 5M Beams, as shown in Figure 7-56.

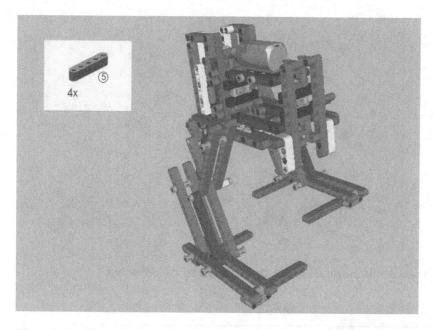

Figure 7-56. *If you want to make a two-legged walker into a four-legged walker, just follow the instructions for Projects 7-1 and 7-2, but use four 5M Beams instead of a second battery box. You won't need another battery box for another pair of legs, just wire the motor to the first battery*

It really isn't difficult to create a pair of legs, the difficult part is syncing them so the creation walks. You might have noticed that at the top of the two-legged walker there is a Z20 Gear. I didn't just include the gear in the build instructions for aesthetic reasons. One thing I have learned is that it is important to make certain the legs are at different levels. So now let's see how to link two pairs of legs together. Figures 7-57 and 7-58 show how to accomplish the final steps to a successful four-legged walking robot.

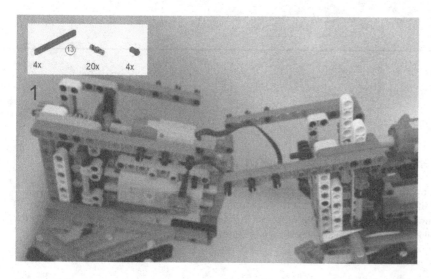

Figure 7-57. *Snap in four Connector Pegs on the top of each pair of legs. Snap in the 13M Beams and insert six Connector Pegs in each one. Snap in four of the Connector Pegs with Bumps on one of the 13M Beams, as shown*

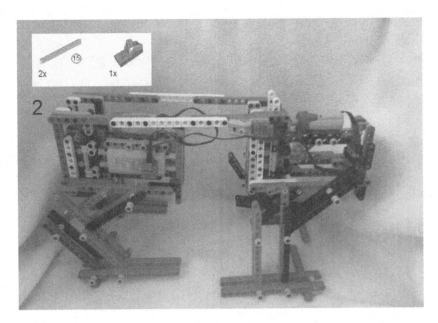

Figure 7-58. *Snap on the 15M Beams, and the two pairs of legs will be locked together. Snap on the switch and use it as an extension for the motor and then connect it to the battery box*

Final Words About Building Robots

Now that you have worked through these seven chapters on building robots on LEGO Technic, I hope you feel you have learned a lot. Now it is your turn. As I have said before, please don't just imitate the models presented in this book. Granted, you can imitate the methods of making an arm, body, or other parts, but the point is that you should use this as a stepping-off point to build a unique robot of your own.

Let me give you a final challenge. You may have noticed that both the two-legged and four-legged walkers can only walk in a straight line. Now, if there was only a way to make them turn? I'm going to leave that up to you to find out!

Summary

When it comes to making robots, it is certainly possible to create them with working legs so they can walk. All it takes is an application of a motor and some legs that will rotate properly so it can take its first steps.

Creating one with two legs is simple, but it is possible to create one with four legs that you can take control of. Any additional legs beyond four is simple once you have the basics, and you could even have a veritable centipede! Have fun as you let your imagination roam with this!

Parts List

The following parts lists are generated by LDView. Part images are provided by Peeron. Circled numbers in the Part columnr epresentp arts izei nn umbero fs tuds.P artc olora ppearsa tt hee ndo fe achd escription.

Parts List for Project 2-1

Part	DesignID	Quantity	Description
	3558	36	Technic 3M Connector Peg with Friction Blue
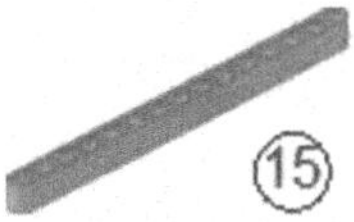	32278	6	Technic 15M Beam Light Gray
	58119	1	BatteryB oxL ightG ray
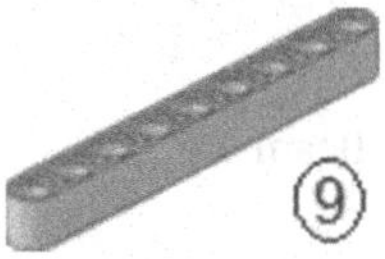	40490	4	Technic 9M Beam Light Gray
	2780	16	Connector Peg with Friction Black
	32526	4	Technic 5 × 3 Angular Beam Light Gray

(continued)

Parts List for Project 2-1 (continued)

Part	DesignID	Quantity	Description
	6562	48	Technic Connector Peg/Cross Axle Blue
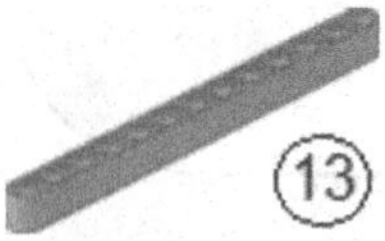	41239	6	Technic 13M Beam Light Gray
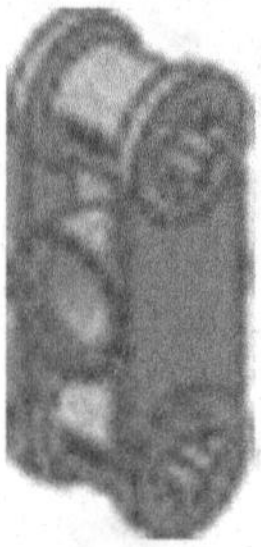	32184	20	Technic Double Cross Block LightG ray
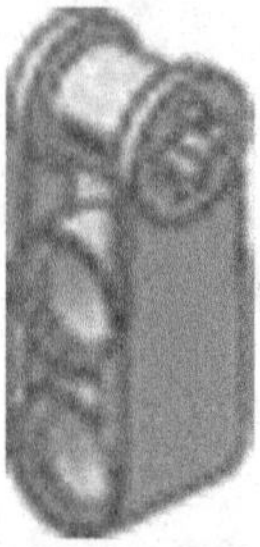	42003	4	Technic 3M Cross Block LightG ray
	32523	4	Technic 3M Beam LightG ray
	60483	4	Technic Cross and Hole Beam LightG ray

Parts List for Project 2-2

Part	DesignID	Quantity	Description
	32140	14	Technic Angular Beam 4 ×2 R ed
	2780	72	Connector Peg with Friction Black
	6562	26	Technic Connector Peg/Cross Axle Blue
	32278	8	Technic15M B eamR ed
	32526	8	Technic 5 × 3 Angular Beam Red
	32525	4	Technic11M B eamR ed
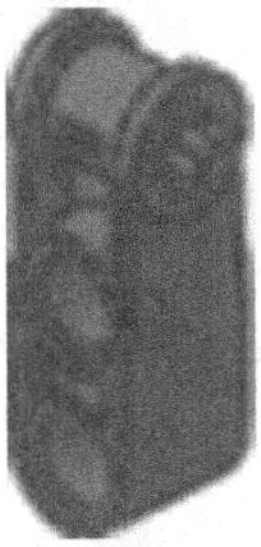	42003	8	Technic 3M Cross Block Red
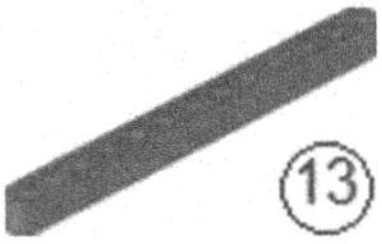	41239	2	Technic13M B eamR ed
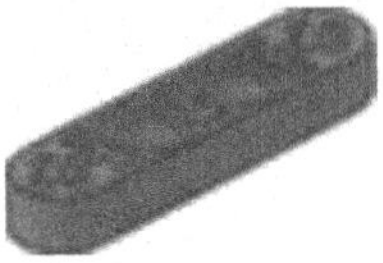	32449	4	Technic4 ML everR ed

(continued)

Parts List for Project 2-2 (*continued*)

Part	DesignID	Quantity	Description
 ④	3705	10	Technic 4M Axle Black
	32013	4	Technic Zero Degree Element (#1) Red
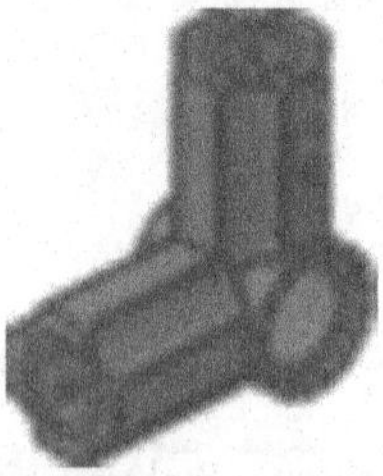	32014	4	Technic 90 Degree Element (#6) Red
	75535	8	TechnicT ube2M R ed
	32034	4	Technic 180 Degree Element Red
	32063	2	Technic6 ML everR ed

Parts List for Project 2-3

Part	DesignID	Quantity	Description
	3558	16	Technic 3M Connector Peg with Friction Blue
	6562	20	Technic Connector Peg/Cross Axle Blue
	32009	4	Technic Double Angular Beam DarkG ray
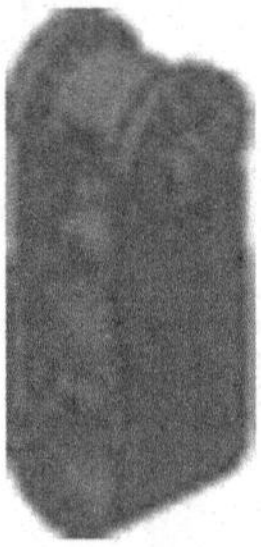	42003	4	Technic 3M Cross Block DarkG ray
	32525	2	Technic 11M Beam Dark Gray
	32526	12	Technic 5 × 3 Angular Beam DarkG ray
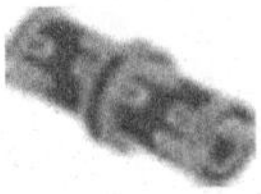	2780	40	Connector Peg with Friction Black
	6629	4	Technic Beam 4 ×6 DarkG ray
	60483	4	Technic Cross and Hole Beam DarkG ray

(*continued*)

Parts List for Project 2-3 (continued)

Part	DesignID	Quantity	Description
	3705	4	Technic 4M Axle Black
	32278	4	Technic 15M Beam DarkG ray
	32523	4	Technic 3M Beam DarkG ray
	6590	4	TechnicB ush LightG ray

Parts List for Project 3-1

Part	DesignID	Quantity	Description
	3701	4	Technic Brick 1 ×4 Blue
	32123	22	Technic Half Bush LightG ray
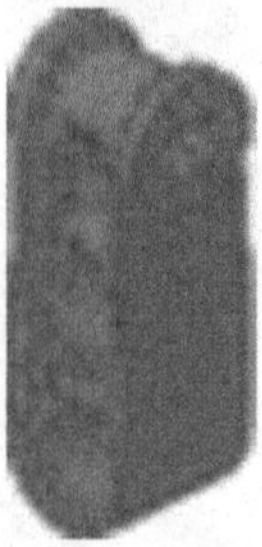	42003	8	Technic 3M Cross Block DarkG ray

(continued)

Parts List for Project 3-1 (*continued*)

Part	DesignID	Quantity	Description
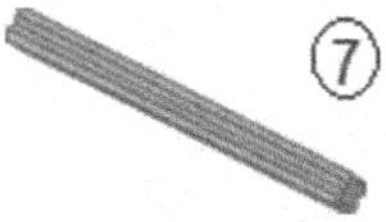 ⑦	44294	6	Technic 7M Axle Light Gray
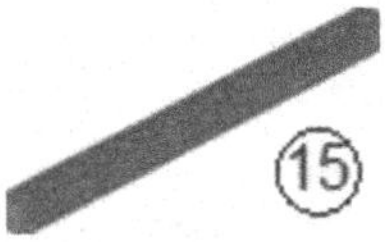 ⑮	32278	8	Technic15M B eamR ed
	2780	22	Connector Peg with Friction Black
	3558	28	Technic 3M Connector Peg with FrictionB lue
	32062	6	Technic 2M Axle with Groove Red
	6536	4	90 Degree Cross Block Red
	6632	8	Technic3 ML everR ed
	58120	2	TechnicM -MotorL ightG ray
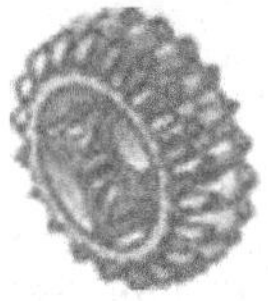	32269	2	Technic Z20 Double Bevel Gear Tan

(*continued*)

Parts List for Project 3-1 (*continued*)

Part	DesignID	Quantity	Description
	32523	2	Technic3M B eamR ed
	6590	12	TechnicB ushL ightG ray
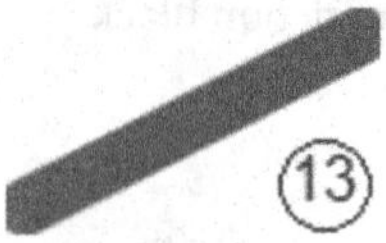	41239	2	Technic13M B eamB lue
	32034	2	Technic 180 Degree Element Blue
	3700	4	Technic Brick 1 ×2 B lue
	32209	2	Technic Axle 5.5 with Stop DarkG ray
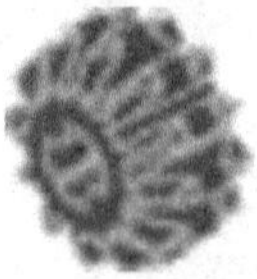	32270	6	Technic Z12 Double Bevel Gear Black
	32316	2	Technic5M B eamR ed
	6562	4	Technic Connector Peg/Cross Axle Blue

(*continued*)

Parts List for Project 3-1 (*continued*)

Part	DesignID	Quantity	Description
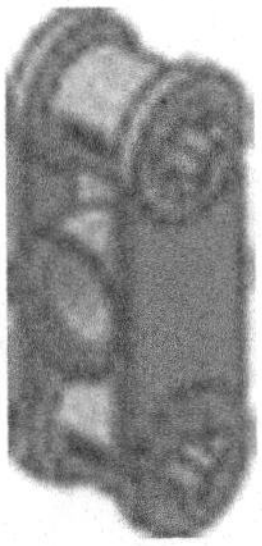	32184	8	Technic Double Cross Block Light Gray
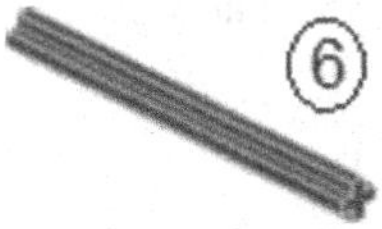 ⑥	3706	2	Technic6 MA xleB lack
 ⑪	32525	2	Technic11M B eamB lue
 ⑨	60485	4	Technic 9M Beam Light Gray
 ④	3705	6	Technic4 MA xleB lack
 ⑨	40490	2	Technic 9M Beam Dark Red
	75535	2	TechnicT ube2M R ed
	32013	2	Technic Zero Degree Element (#1) Blue

Parts List for Project 3-2

Part	DesignID	Quantity	Description
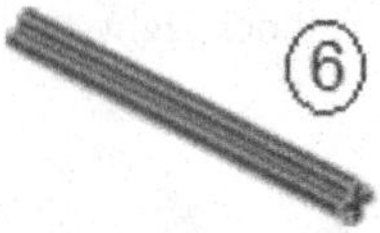 ⑥	3706	8	Technic6 MA xleB lack
	32140	4	Technic Angular Beam 4 ×2 DarkG ray
	6562	8	ConnectorP eg/CrossA xleT an
	6590	35	TechnicB ushL ightG ray
	61903	4	Technic Universal Joint, 3M LightG ray
	60484	4	Technic T-Beam, 3 ×3 D arkG ray
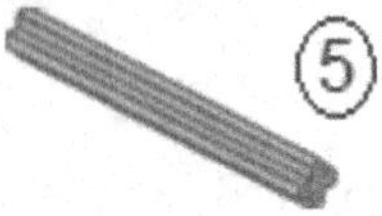 ⑤	32073	4	Technic 5M Axle Light Gray
	63869	4	Cross Block 2 ×3 L ightG ray
	87803	4	Technic 4M Axle with Stop DarkG ray

(continued)

Parts List for Project 3-2 (*continued*)

Part	DesignID	Quantity	Description
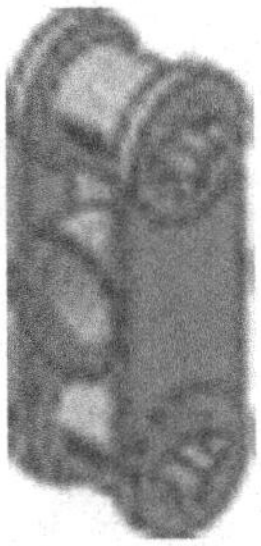	32184	12	Technic Double Cross Block LightG ray
	6587	4	Technic 3M Axle with Bump Tan
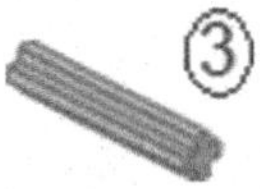	4519	4	Technic 3M Axle Light Gray
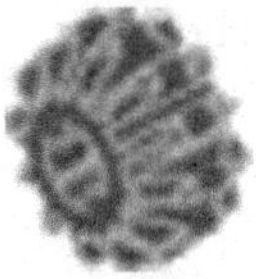	32270	4	Technic Z12 Double Bevel Gear Black
	32269	4	Technic Z20 Double Bevel Gear Tan
	6581	4	Tire 20 ×3 0B lack
	6582	4	RimD arkG ray

(*continued*)

Parts List for Project 3-2 (*continued*)

Part	DesignID	Quantity	Description
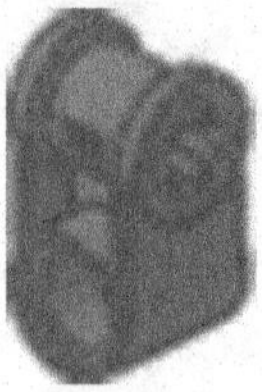	6536	8 .	90 Degree Cross Block Red
	87082	4	Double Bush 3M Light Gray
	3558	4	Technic 3M Connector Peg with FrictionB lue
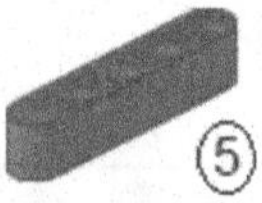 ⑤	32316	4	Technic5M B eamR ed
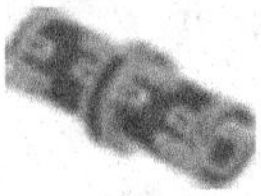	2780	28	Technic Connector Peg with Friction Black
 ⑦	32524	4	Technic7M B eamB lue
	32556	2	Technic 3M Connector Peg Tan
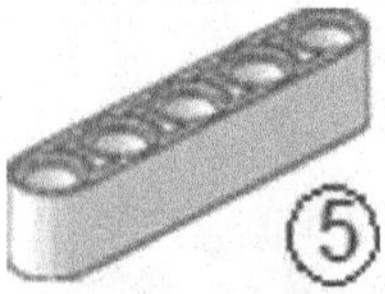 ⑤	32316	4	Technic5M B eamWhite
	32140	4	Technic Angular Beam 4 ×2 B lue
	60483	2	Technic Cross and Hole Beam DarkG ray

(continued)

Parts List for Project 3-2 (*continued*)

Part	DesignID	Quantity	Description
	32526	4	Technic 5 × 3 Angular Beam Blue
	58121	1	TechnicX L-MotorL ightG ray
	3673	4	ConnectorP egL ightG ray
 (11)	32525	2	Technic11M B eamR ed
 (4)	3705	2	Technic4 MA xleB lack
	6562	8	Technic Connector Peg/Cross Axle Blue
	32034	1	Technic 180 Degree Element Blue
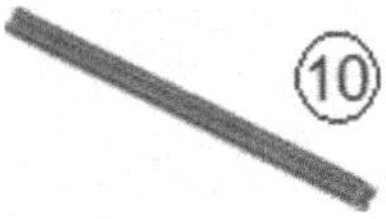 (10)	3737	2	Technic1 0MA xleB lack
	75535	1	TechnicT ube2M R ed
	6538	1	TechnicC ross-AxleE xtensionB lack
	4274	4	Technic Connector Peg with Knob Blue

(*continued*)

Parts List for Project 3-2 (*continued*)

Part	DesignID	Quantity	Description
	3701	3	Technic Brick 1 ×4 R ed
		1	PowerS witchL ightG ray
		1	IR-RXL ightG ray

Parts List for Project 3-3

Part	DesignID	Quantity	Description
	2780	42	Technic Connector Peg with Friction Black
	32140	6	Technic Angular Beam 4 ×2 B lue
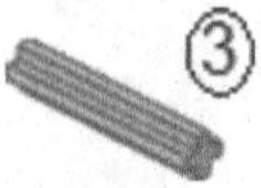	4519	4	Technic3M A xleG ray
	6562	6	Technic Connector Peg/Cross Axle Blue
	32278	2	Technic15M B eamR ed

(*continued*)

Parts List for Project 3-3 (continued)

Part	DesignID	Quantity	Description
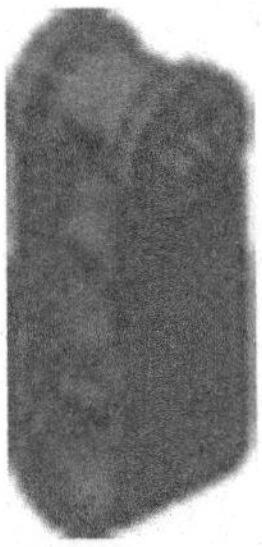	42003	8	Technic 3M Cross Block Dark Gray
	3558	16	Technic 3M Connector Peg with Friction Blue
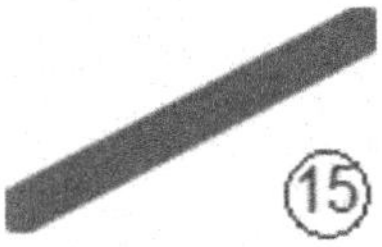	32278	2	Technic15M B eamB lue
	32526	4	Technic 5 × 3 Angular Beam Red
	32523	2	Technic3M B eamR ed
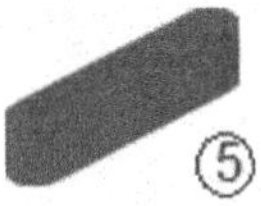	32316	4	Technic5M B eamB lue
	3673	2	ConnectorP egL ightG ray
	32525	2	Technic11M B eamR ed
	76537	4	TechnicS hockA bsorberWhite

Parts List for Project 4-1

Part	DesignID	Quantity	Description
	58120	1	TechnicM -MotorL ightG ray
	2780	9	Technic Connector Peg with Friction Black
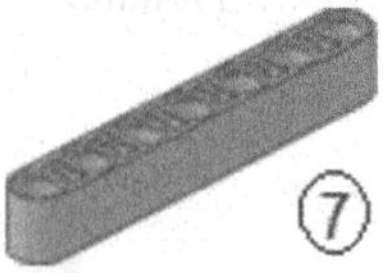 ⑦	32524	3	Technic 7M Beam Light Gray
 ④	3705	2	Technic4 MA xleB lack
	6562	4	Technic Connector Peg/Cross Axle Blue
	32271	4	Technic 3 × 7 Beam Light Gray
 ⑩	3737	1	Technic1 0MA xleB lack
	60483	3	Technic Cross and Hole Beam LightG ray
	6538	2	Technic Cross-Axle Extension LightG ray
	32034	2	Technic 180 Degree Element LightG ray
	75535	1	Technic Tube 2M Light Gray

(continued)

Parts List for Project 4-1 (continued)

Part	DesignID	Quantity	Description
	32062	1	Technic 2M Axle with Groove Red
	3648	2	Technic Gear Wheel 24-tooth LightG ray
	6629	1	Technic Beam 4 ×6 L ightG ray
	32316	1	Technic 5M Beam Light Gray

Parts List for Project 4-2

Part	DesignID	Quantity	Description
	2780	46	Technic Connector Peg with Friction Black
	4519	1	Technic 3M Axle Light Gray
	58121	1	TechnicX L-MotorL ightG ray
	32523	3	Technic3M B eamR ed
	3647	1	Technic Gear Wheel 8-tooth Dark Gray

(continued)

Parts List for Project 4-2 (continued)

Part	DesignID	Quantity	Description
 ⑦	32524	8	Technic7M B eamR ed
	3558	16	Technic 3M Connector Peg with Friction Blue
	50163	1	TechnicT urntable4 .85B lack
	32526	2	Technic 5 × 3 Angular Beam Red
	32523	2	Technic3M B eamB lue
	43857	1	Technic2M B eamR ed
	32000	2	Technic Brick 1 × 2, with two holes Blue
	6562	12	Technic Connector Peg/Cross Axle Blue
	3795	1	Lego Plate, 2 ×6 B lue
	32140	6	Technic Angular Beam 4 ×2 R ed

(*continued*)

Parts List for Project 4-2 (continued)

Part	DesignID	Quantity	Description
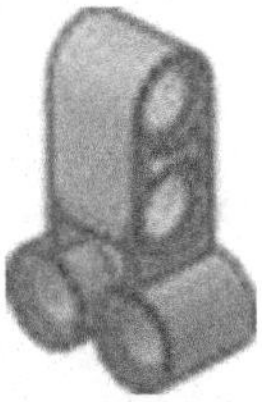	32557	2	Technic Cross Block 2 ×3L ightG ray
	3701	2	Technic Brick 1 ×4 B lue
	32525	4	Technic11M B eamR ed
	32009	4	Technic Double Angular Beam Red
	32316	2	Technic5M B eamR ed

Parts List for Project 4-3

Part	DesignID	Quantity	Description
	2780	82	Technic Connector Peg with Friction Black
	3558	30	Technic 3M Connector Peg with Friction Blue
	6562	44	Technic Connector Peg/Cross Axle Blue
	32523	13	Technic3M B eamWhite
	32449	4	Technic4 ML everW hite

(continued)

Parts List for Project 4-3 (continued)

Part	DesignID	Quantity	Description
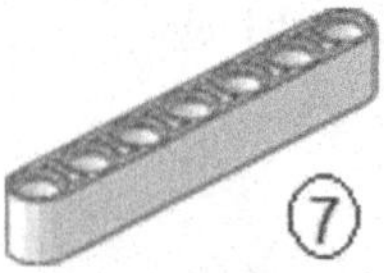	32524	16	Technic7M B eamWhite
	58121	1	TechnicX L-MotorL ightG ray
	32140	6	Technic Angular Beam 4 ×2 White
	32526	16	Technic 5 × 3 Angular Beam White
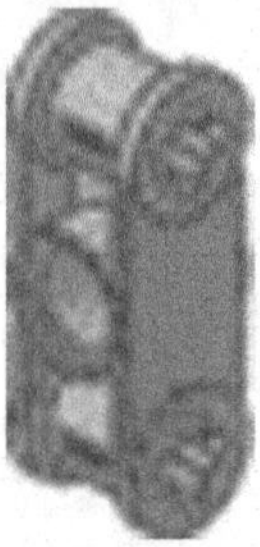	32184	8	Technic Double Cross Block LightG ray
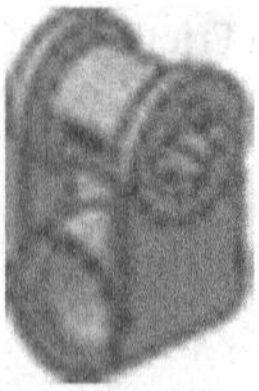	6536	2	90 Degree Cross Block Light Gray
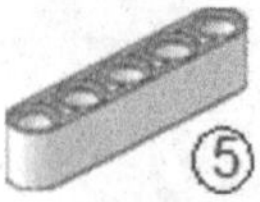	32316	3	Technic5M B eamWhite
	3647	11	Technic Gear Wheel 8-tooth Dark Gray

(continued)

Parts List for Project 4-3 (*continued*)

Part	DesignID	Quantity	Description
	87803	7	Technic 4M Axle with Stop Dark Gray
	6536	16	90 Degree Cross Block White
	3737	4	Technic1 0MA xleB lack
	32291	4	Technic Cross Block 1 ×2 White
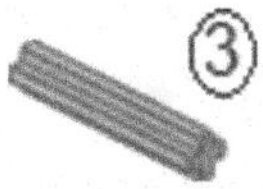	4519	10	Technic 3M Axle Light Gray
	6632	8	Technic3 ML everB lue
	87082	2	Double Bush 3M Light Gray
	32034	1	Technic 180 Degree Element White
	32054	8	Technic Friction Snap Black
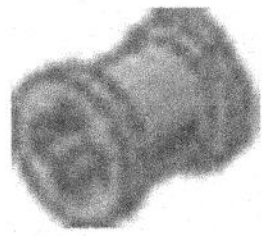	6590	14	TechnicB ushL ightG ray

(*continued*)

Parts List for Project 4-3 (continued)

Part	DesignID	Quantity	Description
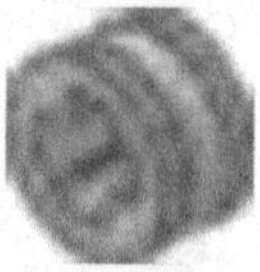	32123	8	Technic Half Bush Light Gray
	6538	2	Technic Cross-Axle Extension Dark Gray
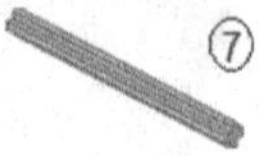 ⑦	44294	1	Technic 7M Axle Light Gray
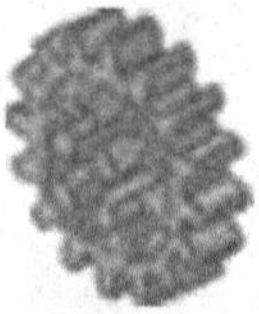	4019	2	Technic Gear 16-tooth Light Gray
	4716	4	Technic Worm Gear Light Gray
 ⑮	32278	4	Technic15M B eamWh ite
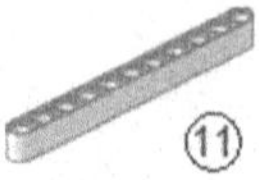 ⑪	32525	4	Technic11M B eamWh ite
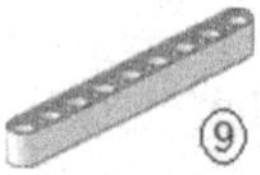 ⑨	40490	4	Technic9M B eamWhite

Parts List for Project 4-4

Part	DesignID	Quantity	Description
 ⑦	32524	6	Technic 7M Beam Dark Gray
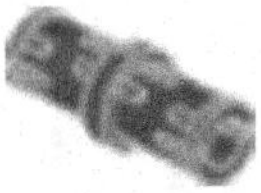	2780	34	Technic Connector Peg with Friction Black
	3558	20	Technic 3M Connector Peg with Friction Blue
	50163	1	TechnicT urntable4 .85B lack
	58121	1	TechnicX L-MotorL ightG ray
	32523	7	Technic3M B eamB lue
	32526	8	Technic 5 × 3 Angular Beam Blue
	6562	31	Technic Connector Peg/Cross Axle Blue
	32140	2	Technic Angular Beam 4 ×2 B lue
 ⑦	32524	6	Technic7M B eamB lue

(continued)

Parts List for Project 4-4 (*continued*)

Part	DesignID	Quantity	Description
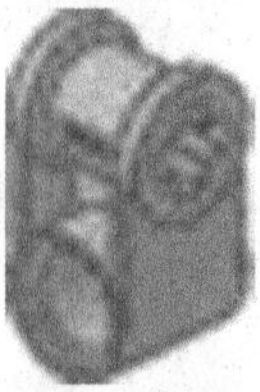	6536	2	90 Degree Cross Block Light Gray
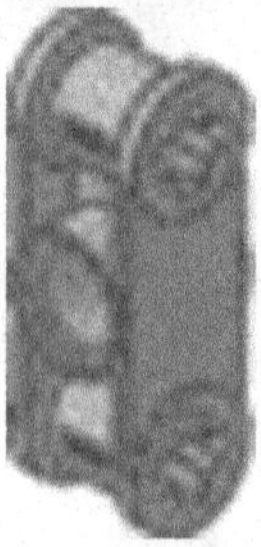	32184	6	Technic Double Cross Block Light Gray
	32316	1	Technic 5M Beam Dark Gray
	32523	2	Technic 3M Beam Dark Gray
	3647	11	Technic Gear Wheel 8-tooth Dark Gray
	87803	7	Technic 4M Axle with Stop Dark Gray
	6536	16	90 Degree Cross Block Blue
	3737	4	Technic1 0MA xleB lack

(*continued*)

Parts List for Project 4-4 (continued)

Part	DesignID	Quantity	Description
	32291	4	Technic Cross Block 1 ×2 B lack
	4519	8	Technic 3M Axle Light Gray
	6632	8	Technic3 ML everB lue
	87082	2	Double Bush 3M Light Gray
	32054	8	TechnicF rictionS napB lack
	6590	14	TechnicB ushL ightG ray
	32123	8	Technic Half Bush Light Gray
	6538	2	Technic Cross-Axle Extension DarkG ray
	32034	1	Technic 180 Degree Element Blue
	44294	1	Technic 7M Axle Light Gray

(*continued*)

Parts List for Project 4-4 (continued)

Part	DesignID	Quantity	Description
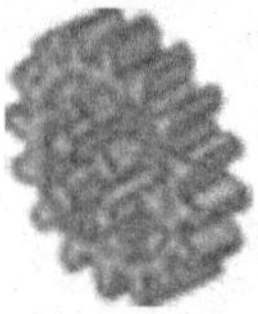	4019	2	Technic Gear 16-tooth Light Gray
	4716	4	Technic Worm Gear Light Gray
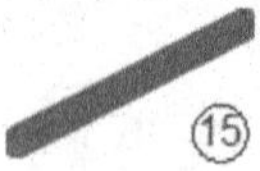	32278	4	Technic15M B eamB lue
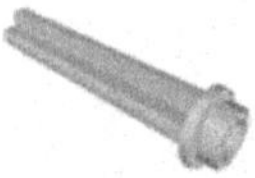	6587	2	Technic Axle 3M with Knob Tan

Parts List for Project 4-5

Part	DesignID	Quantity	Description
	32140	2	Technic Angular Beam 4 ×2 Y ellow
	2780	14	Technic Connector Peg with Friction Black
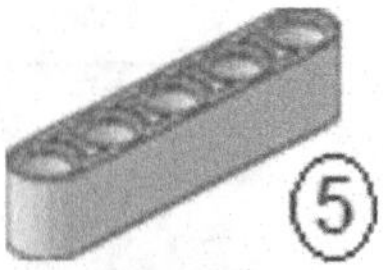	32316	3	Technic5M B eamY ellow
		1	IR-RXL ightG ray
	32064	5	Technic 2 × 1 Brick with Cross-shaped Hole Yellow

(continued)

Parts List for Project 4-5 (continued)

Part	DesignID	Quantity	Description
	58120	1	TechnicM -MotorL ightG ray
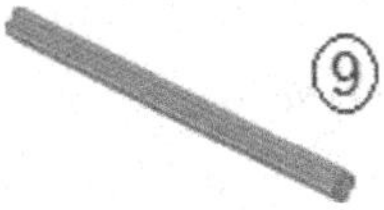	60485	1	Technic 9M Beam Light Gray
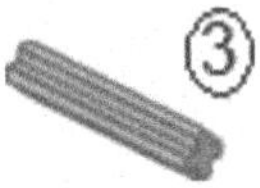	4519	1	Technic 3M Axle Light Gray
	3705	1	Technic4 MA xleB lack
	6562	1	Technic Connector Peg/Cross Axle Tan
	32523	4	Technic3M B eamY ellow
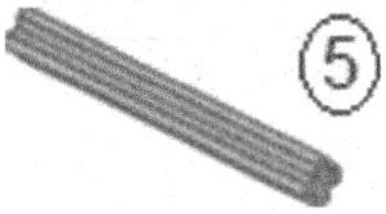	32073	3	Technic5M A xleG ray
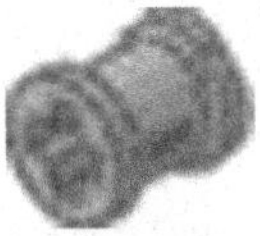	6590	6	TechnicB ushG ray
	3647	3	Technic Gear Wheel 8-tooth Dark Gray
	32062	2	Technic 2M Axle with Groove Red

(continued)

Parts List for Project 4-5 (continued)

Part	DesignID	Quantity	Description
	43093	2	Technic Connector Peg/Cross Axle with Friction Blue
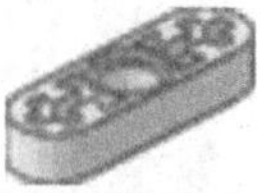	6632	2	Technic3 ML everY ellow
	32013	2	Technic Zero Degree Element (#1) Yellow
	3648	1	Technic Gear Wheel 24-tooth Dark Gray
	3558	2	Technic 3M Connector Peg with Friction Blue
	32184	2	Technic Double Cross Block Light Gray

Parts List for Project 5-1

Part	DesignID	Quantity	Description
	58121	1	Technic XL-Motor Light Gray
	4716	1	TechnicW ormG earB lack
 ⑤	32073	3	Technic 5M Axle Light Gray
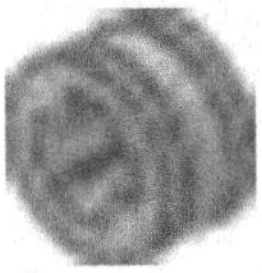	32123	2	Technic Half Bush Light Gray
 ⑦	32524	14	Technic7M B eamR ed
	3558	18	Technic 3M Connector Peg with Friction Blue
	2780	88	Technic Connector Peg with Friction Black
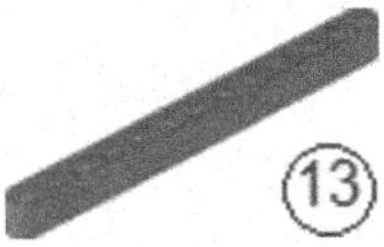 ⑬	41239	12	Technic13M B eamR ed
	32140	8	Technic Angular Beam 4 ×2 R ed
 ⑪	32525	8	Technic11M B eamR ed

(continued)

Parts List for Project 5-1 (continued)

Part	DesignID	Quantity	Description
	6562	28	Technic Connector Peg/Cross Axle Blue
	32523	2	Technic3M B eamR ed
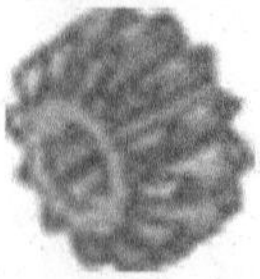	32270	2	Technic Z12 Double Bevel Gear Light Gray
	3647	1	Technic Gear Wheel 8-tooth Dark Gray
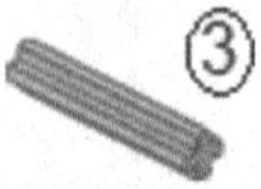	4519	1	Technic 3M Axle Light Gray
	6590	2	TechnicB ushL ightG ray
	6538	2	Technic Cross-Axle Extension Black
	42003	8	Technic 3M Cross Block Dark Gray
	32062	4	Technic 2M Axle with Groove Red
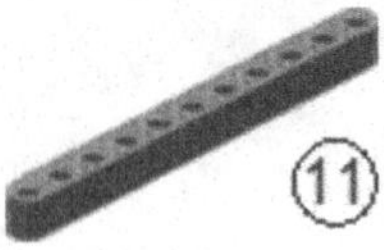	32525	2	Technic1 1MB eamB lack

(*continued*)

Parts List for Project 5-1 (continued)

Part	DesignID	Quantity	Description
	64781	2	Technic1 3MR ackB lack
	32526	6	Technic 5 × 3 Angular Beam Red
	32184	6	Technic Double Cross Block Light Gray
	42003	2	Technic 3M Cross Block Light Gray
	6536	2	90 Degree Cross Block Light Gray
	32316	1	Technic5M B eamR ed

Parts List for Project 5-2

Part	DesignID	Quantity	Description
	58121	1	TechnicX L-MotorL ightG ray
	4716	1	TechnicW ormG earB lack
	32034	5	Technic 180 Degree Element Red
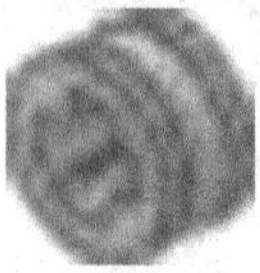	32123	2	Technic Half Bush Light Gray
	6590	7	TechnicB ushL ightG ray
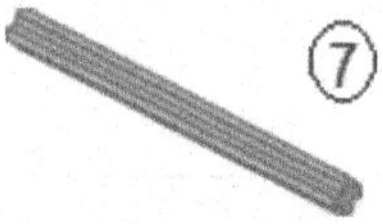 ⑦	44294	2	Technic 7M Axle Light Gray
	6562	28	Technic Connector Peg/Cross Axle Blue
	3558	14	Technic 3M Connector Peg with Friction Blue
	32526	14	Technic 5 × 3 Angular Beam Red
	2780	94	Technic Connector Peg with Friction Black

(continued)

Parts List for Project 5-2 (*continued*)

Part	DesignID	Quantity	Description
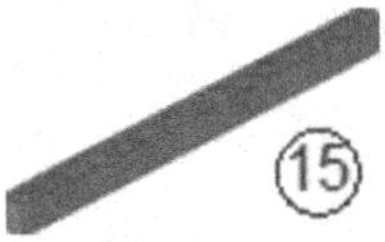 ⑮	32278	18	Technic15M B eamR ed
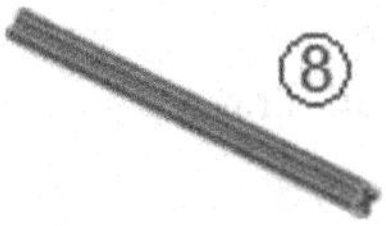 ⑧	3707	2	Technic8 MA xleB lack
	32184	2	Technic Double Cross Block Light Gray
	32523	2	Technic3M B eamR ed
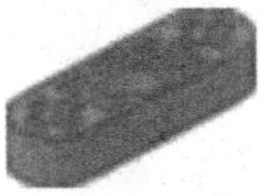	6632	4	Technic3 ML everR ed
	32523	6	Technic3M B eamWhite
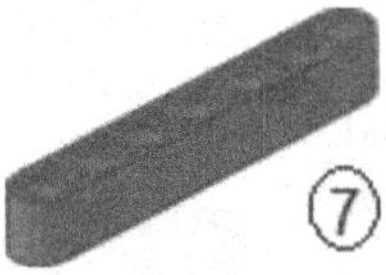 ⑦	32524	6	Technic7M B eamR ed
	6538	6	TechnicC ross-AxleE xtensionB lack
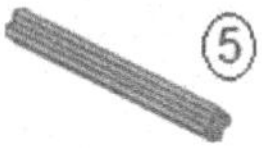 ⑤	32073	2	Technic 5M Axle Light Gray
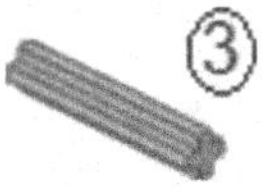 ③	4519	3	Technic 3M Axle Light Gray

(*continued*)

Parts List for Project 5-2 (continued)

Part	DesignID	Quantity	Description
	3648	1	Technic Gear Wheel 24-tooth Dark Gray
	42003	8	Technic 3M Cross Block Dark Gray
	32062	2	Technic 2M Axle with Groove Red
	87761	2	Technic7 MR ackB lack
	32316	4	Technic5M B eamR ed
	32140	10	Technic Angular Beam 4 ×2 R ed
	40490	2	Technic 9M Beam Dark Red
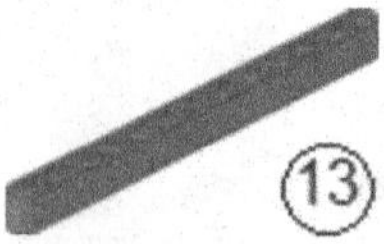	41239	10	Technic13M B eamR ed
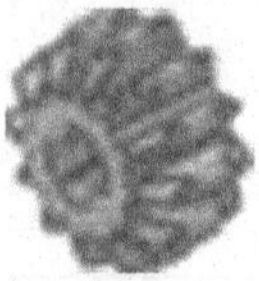	32270	2	Technic Z12 Double Bevel Gear LightG ray

(continued)

Parts List for Project 5-2 (continued)

Part	DesignID	Quantity	Description
	3673	16	ConnectorP egL ightG ray
	75535	2	Technic Tube 2M Light Gray
	6590	2	TechnicB ushL ightG ray
	60483	2	Technic Cross and Hole Beam Black
	3705	4	Technic4 MA xleB lack
	6536	8	90 Degree Cross Block Red

Parts List for Project 6-1

Part	DesignID	Quantity	Description
	32016	16	Technic 157.5 Degree Angle Element (#3) DarkG ray
	6590	14	TechnicB ushB lack
	4519	28	Technic 3M Axle Light Gray
	32054	12	TechnicF rictionS napB lack

(*continued*)

Parts List for Project 6-1 (*continued*)

Part	DesignID	Quantity	Description
⑦	44294	2	Technic 7M Axle Light Gray
	32016	16	Technic 157.5 Degree Angle Element (#3) Red
⑧	3707	4	Technic8 MA xleB lack
	6590	14	TechnicB ushY ellow
	75535	18	Technic Tube 2M Light Gray
	42003	4	Technic 3M Cross Block Red
⑥	3706	14	Technic6 MA xleB lack
	6536	10	90 Degree Cross Block Light Gray
	6538	16	TechnicC ross-AxleE xtensionB lack
	32316	11	Technic5M B eamR ed
⑤	58120	2	TechnicM -MotorL ightG ray

(*continued*)

Parts List for Project 6-1 (*continued*)

Part	DesignID	Quantity	Description
	3558	4	Technic 3M Connector Peg with Friction Blue
	32000	4	Technic Brick 1 × 2, with two holes Blue
	3701	4	Technic Brick 1 ×4 B lue
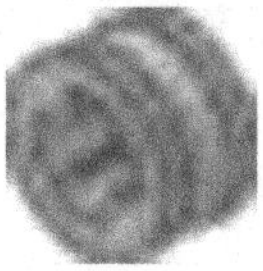	32123	8	TechnicH alfB ushG ray
	32073	2	Technic 5M Axle Light Gray
	32006	2	Technic 4M Lever with Notch Black
	3795	2	Lego Plate, 2 ×6 B lue
	32523	2	Technic3M B eamR ed
	60483	2	Technic Cross and Hole Beam Dark Gray
	2780	2	Technic Connector Peg with Friction Black
	32348	2	Technic 4 ×4 B eamB lack

Parts List for Project 6-2

Part	DesignID	Quantity	Description
	32063	2	Technic6 ML everB lue
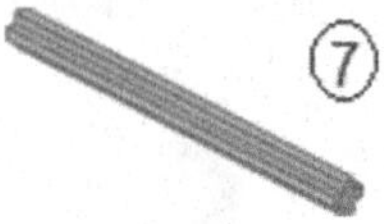	44294	3	Technic 7M Axle Light Gray
	6590	10	TechnicB ushL ightG ray
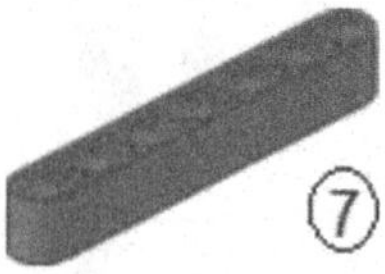	32524	2	Technic7M B eamR ed
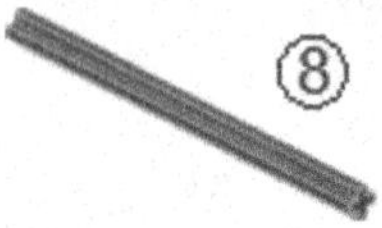	3707	16	Technic8 MA xleB lack
	6536	2	90 Degree Cross Block Red
	32013	2	Technic Zero Degree Element (#1) Red
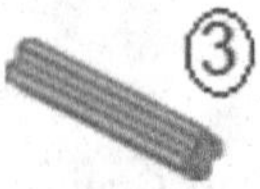	4519	15	Technic 3M Axle Light Gray
	75535	14	Technic Tube 2M Light Gray
	32073	1	Technic 5M Axle Light Gray
	32316	1	Technic5M B eamB lue

(continued)

Parts List for Project 6-2 (*continued*)

Part	DesignID	Quantity	Description
	32140	2	Technic Angular Beam 4 ×2 B lue
	2780	8	Technic Connector Peg with Friction Black
	32316	20	Technic5M B eamR ed
	32123	10	Technic Half Bush Light Gray
	3705	1	Technic4 MA xleB lack
	32062	20	Technic 2M Axle with Groove Red
	41677	4	Technic2 ML everR ed
	32013	4	Technic Zero Degree Element (#1) LightG ray
	42003	4	Technic 3M Cross Block Red
	6632	4	Technic3 ML everR ed
	32192	16	Technic 135 Degree Angle Element (#4) Yellow

(*continued*)

Parts List for Project 6-2 (continued)

Part	DesignID	Quantity	Description
	32054	16	TechnicF rictionS napB lack
	6536	2	90 Degree Cross Block Light Gray
	32184	2	Technic Double Cross Block Black
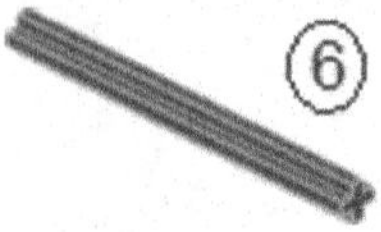 (6)	3706	4	Technic6 MA xleB lack
	58119	1	BatteryB oxL ightG ray
	6590	14	TechnicB ushB lack
	32016	16	Technic 157.5 Degree Angle Element (#3) DarkG ray
 (9)	40490	2	Technic 9M Beam Dark Gray

Parts List for Project 6-3

Part	DesignID	Quantity	Description
	3558	24	Technic 3M Connector Peg with Friction Blue
	40490	8	Technic 9M Beam Dark Red
	32523	2	Technic3M B eamR ed
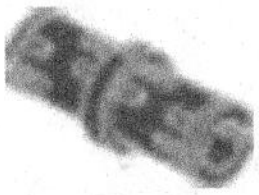	2780	44	Technic Connector Peg with Friction Black
		2	IR-RXL ightG ray
	32526	8	Technic 5 × 3 Angular Beam Blue
	32278	4	Technic15M B eamR ed
	32526	4	Technic 5 × 3 Angular Beam Light Gray
	60484	1	Technic T-Beam, 3 ×3 D arkG ray
	6562	2	Technic Connector Peg/Cross Axle Blue

(continued)

Parts List for Project 6-3 (*continued*)

Part	DesignID	Quantity	Description
	32184	1	Technic Double Cross Block Light Gray
	58121	1	TechnicX L-MotorL ightG ray
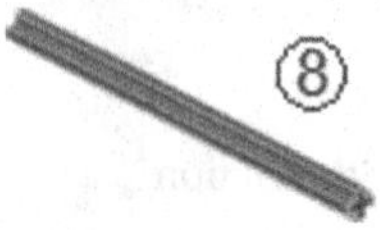	3707	1	Technic8 MA xleB lack
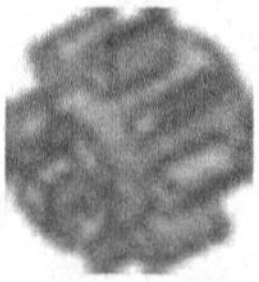	3647	2	Technic Gear Wheel 8-tooth Light Gray
	32062	7	Technic 2M Axle with Groove Red
	32073	3	Technic 5M Axle Light Gray
	6590	7	TechnicB ushL ightG ray
	75535	2	Technic Tube 2M Light Gray
	32013	2	Technic Zero Degree Element (#1) Blue
	32016	8	Technic 157.5 Degree Angle Element (#3) Blue

(*continued*)

Parts List for Project 6-3 (*continued*)

Part	DesignID	Quantity	Description
	6538	2	TechnicC ross-AxleE xtensionB lack
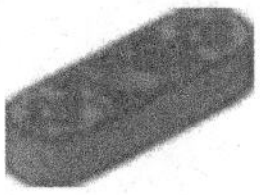	6632	4	Technic3 ML everR ed
	41239	1	Technic 13M Beam Dark Gray
	4519	1	Technic 3M Axle Light Gray
	3708	1	Technic1 2MA xleB lack

Parts List for Project 7-1

Part	DesignID	Quantity	Description
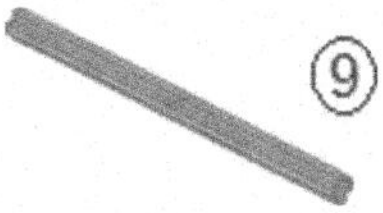	60485	2	Technic 9M Beam Light Gray
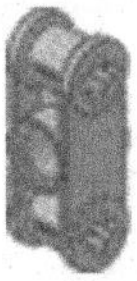	32184	2	Technic Double Cross Block Light Gray
	58119	1	BatteryB oxL ightG ray
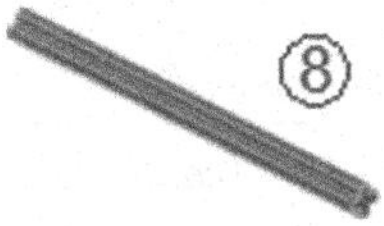	3707	2	Technic8 MA xleB lack

(*continued*)

Parts List for Project 7-1 (*continued*)

Part	DesignID	Quantity	Description
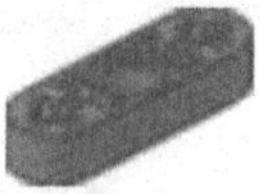	6632	6	Technic3 ML everR ed
	42003	4	Technic 3M Cross Block Dark Gray
	2780	74	Technic Connector Peg with Friction Black
	6536	12	90 Degree Cross Block Dark Gray
	40490	6	Technic 9M Beam Dark Red
	2817	2	Technic Bearing Plate 2 ×2 Black
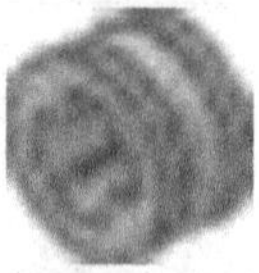	32123	8	Technic Half Bush Light Gray
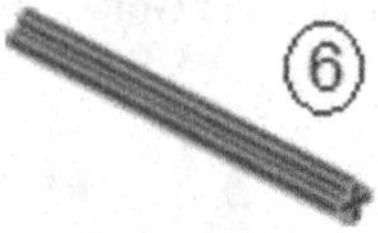	3706	2	Technic6 MA xleB lack
	32034	2	Technic 180 Degree Element Red
	6562	10	Technic Connector Peg/Cross Axle Blue

(*continued*)

Parts List for Project 7-1 (continued)

Part	DesignID	Quantity	Description
	75535	1	Technic Tube 2M Light Gray
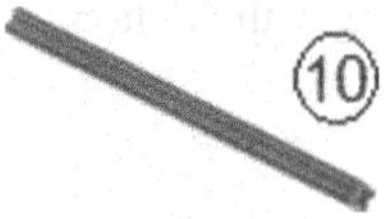 ⑩	3737	1	Technic1 0MA xleB lack
	58120	1	TechnicM -MotorL ightG ray
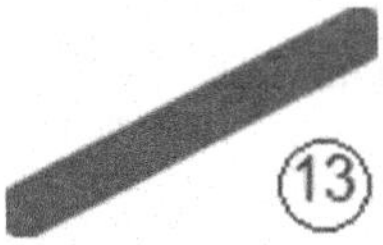 ⑬	41239	2	Technic13M B eamB lue
	32291	4	Technic Cross Block 1 ×2 Black
	32123	2	TechnicH alfB ushY ellow
	4716	1	Technic Worm Gear Light Gray
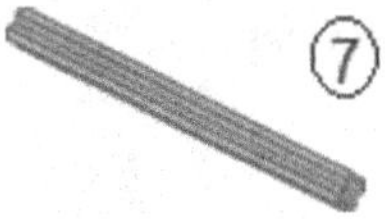 ⑦	44294	2	Technic 7M Axle Light Gray
	32123	4	TechnicH alfB ushB lack
	75535	2	TechnicT ube2M R ed

(continued)

Parts List for Project 7-1 (continued)

Part	DesignID	Quantity	Description
	32013	4	Technic Zero Degree Element (#1) Red
	3558	16	Technic 3M Connector Peg with Friction Blue
	32140	8	Technic Angular Beam 4 ×2 B lue
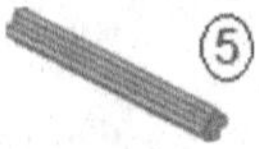 ⑤	32073	4	Technic 5M Axle Light Gray
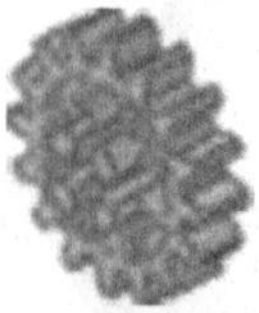	4019	4	Technic Gear 16-tooth Light Gray
	6590	6	TechnicB ushB lack
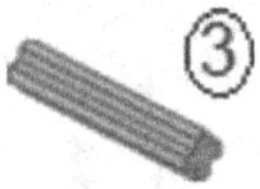 ③	4519	4	Technic 3M Axle Light Gray
 ⑦	32524	3	Technic7M B eamR ed
	32269	1	Technic Z20 Double Bevel Gear LightG ray
	6590	1	TechnicB ushL ightG ray

(continued)

Parts List for Project 7-1 (continued)

Part	DesignID	Quantity	Description
	32140	6	Technic Angular Beam 4 ×2 White
	32526	4	Technic 5 × 3 Angular Beam Red
	32525	4	Technic11M B eamR ed
	32524	2	Technic7M B eamWhite
	6538	2	Technic Cross Axle Extension Black
	32316	2	Technic5M B eamB lue
	32524	2	Technic7M B eamB lue
	32278	2	Technic15M B eamR ed
	3705	2	Technic4 MA xleB lack

Parts List for Project 7-2

Part	DesignID	Quantity	Description
 ⑥	3706	10	Technic6 MA xleB lack
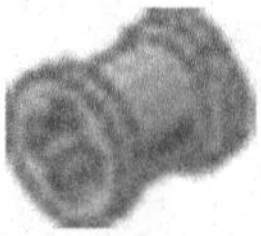	6590	16	TechnicB ushL ightG ray
	6632	4	Technic3 ML everR ed
	32556	2	Technic 3M Connector Peg Tan
 ⑪	32525	6	Technic11M B eamR ed
 ⑨	40490	8	Technic 9M Beam Dark Red
	6562	2	ConnectorP eg/CrossA xleT an
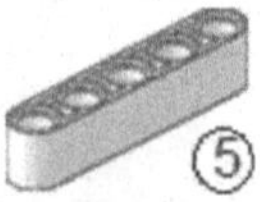 ⑤	32316	2	Technic5M B eamWhite
	32123	8	Technic Half Bush Light Gray
 ⑮	32278	6	Technic15M B eamR ed

(continued)

Parts List for Project 7-2 (continued)

Part	DesignID	Quantity	Description
	63869	4	Cross Block 2 ×3 L ightG ray
	4519	4	Technic 3M Axle Light Gray
	2780	8	Technic Connector Peg with Friction Black

Index

■ P, Q

■ S, T, U, V

■ W, X, Y, Z